Applied Wavelet
Analysis with S-PLUS

Andrew Bruce   Hong-Ye Gao

Applied Wavelet
Analysis with S-PLUS

With 192 illustrations

Springer

Andrew Bruce
MathSoft, Inc.
1700 Westlake Avenue North
Suite 500
Seattle, WA 98109-9891 USA

Hong-Ye Gao
MathSoft, Inc.
1700 Westlake Avenue North
Suite 500
Seattle, WA 98109-9891 USA

Library of Congress Cataloging-in-Publication Data
Bruce, Andrew
    Applied wavelet analysis with S-PLUS / Andrew Bruce, Hong-Ye Gao
      p.   cm.
    Includes bibliographical references and index.
    ISBN 0-387-94714-0 (softcover : alk. paper)
      1. Signal processing—Mathematics. 2. Wavelets (Mathematics) –
    –Data processing. 3. S-PLUS (Computer program language). 4. Computer
    –aided engineering. I. Gao, Hong-Ye. II. Title.
    TK5102.9.B78   1996
    621.382'2'015152433—dc20   96-12971

Printed on acid-free paper.

Printed in the United States of America.   (MVY)

9 8 7 6

ISBN 0-387-94714-0

Springer-Verlag is a part of *Springer Science+Business Media*

springeronline.com

Preface

Wavelets are the topic of much excitement in mathematics, engineering, and scientific circles. Wavelet theory has unified and extended ideas from several domains, including subband filtering, approximation theory, signal processing, image processing, and nonparametric estimation. Not just researchers are using wavelets. Wavelets are being applied to a diverse set of problems, and have the potential to make significant technological advancements in fields ranging from consumer electronics to wireless communications to medical diagnostic imaging.

Wavelets and related analytical techniques have been the subject of a number of books and review articles. Most of these books are mathematical in nature, requiring an advanced degree in the mathematical sciences or in engineering. This book is intended for use by a broad range of data analysts, scientists, and engineers. While readers should be familiar with calculus and linear algebra at the undergraduate level, reliance on formal mathematical methods is minimized.

This book introduces applied wavelet analysis through a visual data analysis approach based on the wavelets module of the S-PLUS software system. Wavelet concepts are explained in a way that is intuitive and easy to understand. Wavelets, as well as a whole range of related signal processing techniques, such as wavelet packets, local cosine analysis, and matching pursuits, are covered. Applications of wavelet analysis are illustrated, including nonparametric function estimation, digital image compression, and time-frequency acoustic signal analysis.

In chapter 1, we illustrate some of the applications of wavelet analysis, including signal de-noising, time-frequency analysis, digital image compression, and signal classification. In chapters 2–4, we present the basics of one-dimensional and two-dimensionsal wavelet analysis. Chapter 2 covers several fundamental topics, including the wavelet approximation, the discrete wavelet transform (DWT), the inverse transform (IDWT), multiresolution analysis, time-scale analysis, and the pyramid algorithm. Chapter 3 extends the concepts discussed in chapter 2 to analysis of two-dimensional image data. Using graphical and statistical summaries, chapter 4 illustrates some key properties of wavelet analysis, including the "energy compression" property.

Chapter 5 discusses some options in wavelet analysis, and gives details on using the S+WAVELETS software for wavelet analysis. While this chapter is important for those who want to program in S+WAVELETS, it can be skipped in a first reading. Chapter 6 explains how to use wavelets for nonparametric estimation using the Donoho and Johnstone WaveShrink technique. WaveShrink is a very significant development in statistical theory and is proving valuable in practice.

Chapters 7–9 describe more general families of time-frequency representations. These families are an important extension to the original wavelet approach. Chapter 7 describes wavelet packets, which include wavelets as a special case. The Coifman and Wickerhauser best basis algorithm for adaptive basis selection is discussed in this chapter. Chapter 8 describes cosine packet analysis, also known as local cosine analysis. Cosine packet analysis involves a family of time-frequency decompositions through special tapers applied to trigonometric functions. Many of the concepts for wavelet packet analysis, such as best basis selection, apply to cosine packet analysis. Chapter 9 extends wavelet packet and cosine packet analysis to two dimensional image data, and gives an extended example involving a digital fingerprint image and the proposed FBI standard for compression.

Chapters 10–11 discuss a variety of other related signal processing techniques. Chapter 10 is devoted to the Mallat and Zhang matching pursuit algorithm and other "molecular" decompositions. Chapter 11 describes non-decimated wavelet analysis, the à trous algorithm, and the outlier resistant wavelet analysis. The non-decimated wavelet transform, also known as the translation-invariant or stationary wavelet transform, has proven especially valuable in applications such as signal de-noising. Also in chapter 11 is a section on creating and using new wavelets.

Chapters 12–14 give the technical details for the algorithms behind wavelet, wavelet packet, and cosine packet analysis. Chapter 12 describes the basic wavelet filter algorithms and formulas. Several different viewpoints are adopted, including a signal processing approach, a matrix algebra approach, and a function approximation approach. Chapter 13 gives the basic algorithms for wavelet packets and cosine packets and chapter 14 dicusses the treatment of the boundaries.

Comments?

We welcome your comments on this book. Please send electronic mail to following addresses:

```
andrew@statsci.com
gao@statsci.com
```

Acknowledgements:

The research and development leading to this book was partially supported by the NASA Small Business Innovation Research (SBIR) Contract No. NAS13-587. The technical monitor for this contract is Bruce Davis at the John C. Stennis Space Center, Mississippi.

This book and the S+WAVELETS software are the product of efforts by many others at StatSci, the University of Washington, and Stanford University. We are particularly indebted to Professor R. Douglas Martin (StatSci and University of Washington) and Professor David L. Donoho (Stanford University). Prof. Martin started the wavelets project at StatSci and was instrumental in seeing that S+WAVELETS move from a research project to a commercial product. He also made substantial contributions to chapters 2 and 12. Prof. Donoho laid the foundations for the approach and style adopted in our book and software. Many of the ideas and examples have been taken directly from his work.

The manual was edited and formatted by the StatSci documentation group. Richard Calaway was the primary editor of the original version of this book. He provided extensive advice and feedback on both the technical content and stylistic nature of the book. Kjrsten Henriksen also edited the original version, and was responsible for building the index. John Minardi edited and formatted the current version of this book. Lisa Eaton (StatSci) and Tanya Friedman (Rikki Conrad Design) produced the cover.

The S+WAVELETS software module was developed by the authors. In addition to Prof. Donoho, several consultants made major contributions to the design of the toolkit. David Ragozin (University of Washington) was the key "hands-on" consultant, developing wavelet algorithms and making many important contributions to the design. Eve Riskin and Jill Goldschneider (University of Washington) provided valuable input to the design and testing of the toolkit.

Significant contributions to the toolkit were made by other members of the StatSci wavelets research team. Emmanuel Arbogast was involved in source code administration, porting, C code development, and frequent firefighting on many fronts. M. Y. Jaisimha helped in source code administration and porting, developed C code for wavelet analysis, and also fought many fires. Heather Van Steenburgh was in charge of porting and release engineering for the toolkit. Luke Goodwin was in charge of QA testing and beta test site administration.

Mary Ellen Bock (Purdue University), Jon Buckheit (Stanford University), Marianna Clark (StatSci), Doug Clarkson (StatSci), Gary Hewer (China Lake Naval Weapons Center), Wei Kuo (China Lake Naval Weapons Center), Iain Johnstone (Stanford University), Charles Roosen (StatSci) along with many other readers and users, provided many helpful comments on early versions of this book and S+WAVELETS.

Andrew Bruce
Hong-Ye Gao

Contents

S-PLUS and S+WAVELETS

The S-PLUS software system is a flexible, graphical environment for visually exploring and modeling your data. Using graphical displays and other data analysis tools, you can obtain an understanding of the structure of your data, determine distributions, find relationships, and much more. With a built-in object-oriented language, S-PLUS is a powerful scientific computing environment. S-PLUS offers a rich collection of classical and modern statistical modeling techniques for advanced exploratory and confirmatory analysis.

The S+WAVELETS module is an extension to the S-PLUS language. S+WAVELETS offers an object-oriented language and a comprehensive computing environment for wavelet analysis. A variety of wavelet representations are available, along with tools to visualize, manipulate, synthesize, and analyze these representations.

S-PLUS and S+WAVELETS are products of the StatSci division of MathSoft, Inc., and are available for UNIX and Microsoft Windows. For more information, contact StatSci in any of the following ways:

- E-mail to mktg@statsci.com.

- Fax to **(206) 283-8691**.

- Call **(800) 569-0123** or **(206) 283-8802**, Monday through Friday between 7:30 a.m. and 5:00 p.m. Pacific time.

Typographic Conventions

This book obeys the following typographic conventions:

- The *italic font* is used for emphasis, and also for user-supplied variables within UNIX, DOS, and S-PLUS commands. For example,

  All objects have implicit, defining, and optional *attributes*.

- The `typewriter font` is used for S-PLUS functions and examples of S-PLUS sessions. For example,

  ```
  > plot(corn.rain)
  ```

  Displayed S-PLUS commands are shown with the prompt `>`. S-PLUS commands that require more than one line of input are displayed with the continuation prompts indicated by `+` or `Continue string:`.

 Warning: When you see the "dangerous bend" sign followed by the word **Warning**, you are seeing a warning about S-PLUS behavior. Read these warnings carefully.

Note: Points of interest are preceded by the word **Note**.

The S-PLUS output and plots were generated from S-PLUS Version 3.3 and S+WAVELETS Version 1.1. The S-PLUS output was generated with the options setting

```
options(width=60)
```

Some of the output has been hand-edited to avoid line overflow.

Terminology

Atom: A coefficient corresponding to either a single wavelet, a wavelet packet, or a cosine packet.

Best basis algorithm: An algorithm developed by Coifman and Wickerhauser to adaptively select an orthogonal transform from a wavelet packet or cosine packet table. The selected transform optimizes an additive costs criterion for an observed signal or image.

Block DCT: The discrete cosine transform applied, without tapering, to equally sized blocks of a signal or image.

Block CPT: The cosine packet transform applied to equally sized blocks of a signal or image.

Cosine packet: A tapered sinusoid where the taper is chosen to ensure orthogonality between sinusoids in adjacent blocks.

CPT: The cosine packet transform which involves applying a tapered discrete cosine transform to signal or image blocks. Special tapers are used to ensure orthogonality between adjacent blocks. This is also known as the local cosine transform.

CP table: A complete enumeration of cosine packet crystals up to a specified temporal/spatial resolution level. The crystals are laid out in tabular form convenient for selecting orthogonal transforms using, for example, the best basis algorithm.

Crystal: There are two types of crystals: wavelet/wavelet packet crystals and cosine packet crystals. Wavelet and wavelet packet crystals are indexed by location, and cosine packet crystals are indexed by frequency. A wavelet packet crystal is commonly referred to as a *subband* in the engineering and signal processing community. The terminology crystal derives from the fact that crystals are collections of atoms organized on a lattice.

DCT: The discrete cosine transform without tapering.

Dictionary: A dictionary contains all the information necessary for analysis and synthesis of a signal or image. A dictionary includes information such as the wavelet, transform type, and boundary rule.

DWT: The discrete wavelet transform.

EDA: Exploratory data analysis, which is a collection of visual techniques for exploring data.

Matching-pursuits algorithm: An algorithm developed by Mallat and Zhang for obtaining time-frequency representations from arbitrary collections of waveforms. A matching-pursuit representation is typically not orthogonal and produces a molecule instead of a crystal.

Molecule: A vector of coefficients corresponding to an unorganized collection of atoms. In contrast to a crystal, the atoms are not indexed by location or frequency.

Multiresolution analysis: Decomposition of a signal into components at different resolution levels. The fine and coarse resolution components capture, respectively, the fine and coarse scale features in the signal.

Non-decimated DWT: An overcomplete discrete wavelet transform with the same number of coefficients at each resolution level. The non-decimated DWT is also known as the translation-invariant wavelet transform or the stationary wavelet transform.

Smoother-cleaner DWT: A fast wavelet decomposition which is robust towards outliers.

WaveShrink: The Donoho and Johnstone wavelet shrinkage procedure for nonparametric estimation of signals and functions from noisy data.

WPT: A wavelet packet transform which involves wavelet filters applied to a signal or image to create a collection of wavelet packet crystals (subbands) at different frequencies.

WP table: A complete enumeration of wavelet packet crystals (subbands) up to a specified frequency resolution level. The crystals are laid out in tabular form convenient for selecting orthogonal transforms using, for example, the best basis algorithm.

1
Introduction

Wavelets are being applied in a diverse set of fields, such as signal processing, medical imaging, pattern recognition, data compression, and numerical analysis. Wavelet theory has inspired the development of a powerful metholodogy for processing signals, images, and other types of scientific and technical data. This methodology includes a wide range of tools, such as the wavelet transform, multiresolution analysis, time-scale analysis, time-frequency representations with wavelet packets and cosine packets, "best basis" analysis, "matching pursuit" decompositions, overcomplete representations with non-decimated wavelets, and outlier resistant wavelet transforms.

The study of wavelets as a distinct discipline started in the late 1980's. The story of wavelets, however, is only just beginning. The applications mentioned in this book are just a small sample, and many more examples can be cited. Wavelet research is proceeding at a rapid pace. Significant new developments appear each year, vastly expanding the domain of wavelet analysis. In a few years' time, the use of wavelet analysis may become as ubiquitous as the use of Fourier analysis.

In this book, we present the essentials of applied wavelet analysis through the S+WAVELETS toolkit, a fully supported commercial module of the S-PLUS software environment. S+WAVELETS is an interactive object-oriented language and software system for wavelet analysis of signals and images. It offers a full complement of wavelet and visual display tools. The object-oriented nature of the language gives an organized framework for wavelet analysis and provides a language interface that is both easy to use and readily customizable for specific applications.

In the remainder of this chapter, we present four wavelet applications: digital image compression, noise removal, time-frequency analysis, and the speeding up and improvement of classification algorithms. These applications can all be implemented using high-level S+WAVELETS functions. We conclude the chapter with some references to other books and resources for learning more about wavelet analysis.

1.1   Example 1: Digital Image Compression

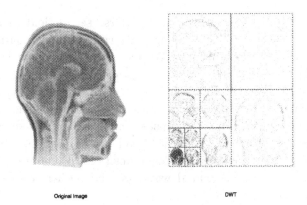

Original Image   DWT

FIGURE 1.1. A magnetic resonance (MR) image of the brain (left), and its wavelet transform with three multiresolution levels (right). The wavelet transform is very sparse compared with the original image: most of the wavelet coefficients are close to zero.

The amount of digital image data which is being stored and transmitted has exploded in the past decade. The types of data range from medical images to still photography to document images to fingerprint images. Compression algorithms and software are needed to deal with this proliferation of data. A very important class of algorithms for image compression are based on *transform coding*. Rather than encoding an image directly, the image is first transformed and the transform coefficients are then encoded.

Figure 1.1 displays a magnetic resonance (MR) image of the brain of Eve Riskin. Professor Riskin is a leading researcher in data compression at the University of Washington. Plotted alongside the brain image are the absolute values of the wavelet transform coefficients. The darkness of the pixels corresponds to the magnitude of the coefficients. The wavelet transform decomposes the original image into waveforms at different scales and locations. The large wavelet coefficients tend to correspond to important features in the image, such as edges. Figure 1.1 shows that the wavelet transform is very sparse compared with the original image: most of the wavelet coefficients are close to zero. The sparsity of the wavelet transform is the key for compression efficiency. A very sparse transform, which has fewer "important" coefficients to encode, can be efficiently compressed.

The current standard for compression of digital images is the JPEG (Joint Photographics Experts Group) transform coding algorithm. JPEG is commonly available in many commercial software and imaging systems. JPEG encodes coefficients from the discrete cosine transform (DCT) applied to 8×8 pixel blocks of the image. For many images the wavelet transform tends to be more sparse than the DCT, and hence can lead to lower distortions at higher compression ratios. The wavelet transform, which can efficiently represent edges, also leads to fewer artifacts in the compressed image.

One of the most celebrated applications of wavelets is the proposed FBI standard for compression of digital fingerprint data [Cri93]. This standard is based on a wavelet packet transform. A digital fingerprint is captured at a resolution of 500 pixels per inch and 256 gray scale levels (8 bits). This means that a typical uncompressed fingerprint image will require roughly half a megabyte of storage. Figure 1.2 displays a portion of a digital fingerprint image along with reconstructions based on the top five percent of the coefficients of three different transforms: the DCT, the discrete wavelet transform, and wavelet packet transform used in the FBI standard. The DCT reconstruction is highly distorted and exhibits serious "blocking" artifacts. The wavelet packet reconstruction is visually the most appealing. For more details about digital fingerprint compression, refer to [BBH93, Hop94].

Original Image   DWT

Block DCT   Wavelet Packet

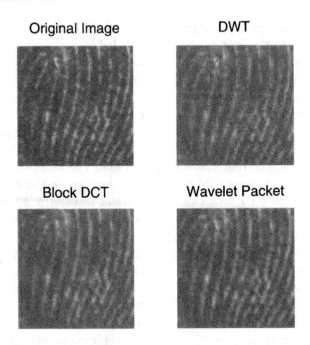

FIGURE 1.2. Wavelets have been successfully used in image compression problems. Pictured above is a 128 by 128 pixel block of a digital fingerprint with 256 gray-levels (upper left). Also pictured are reconstructions based on the top five percent of the coefficients from three different transforms: the wavelet transform (upper right), the block DCT (lower left), and a wavelet packet transform (lower right). The wavelet reconstructions are visually more appealing than the DCT.

1.2   Example 2: Noise Removal

Statistical prediction or function estimation is of fundamental importance. For many prediction problems, the most appropriate approach is a nonparametric procedure. Wavelet estimation methodology, pioneered by Donoho and Johnstone [Don93a, DJ94, DJKP95, Don95, DJ95], represents an important breakthrough in terms of both the theory and the practice of nonparametric estimation and prediction. Donoho and Johnstone show that wavelet estimation has very broad asymptotic near-optimality properties. These properties are derived, in large part, from the ability of wavelets to represent locally nonsmooth phenomena with relatively few coefficients.

Donoho and Johnstone developed the WaveShrink procedure for

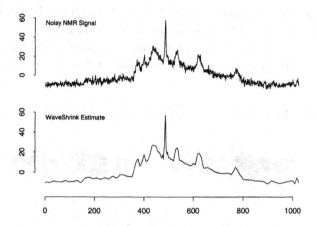

FIGURE 1.3. A noisy nuclear magnetic resonance (NMR) signal with 1024 sample values (top panel) and the WaveShrink estimate of the signal (bottom panel). An important feature of WaveShrink is its ability to remove noise while preserving non-smooth features, such as the large spike in the NMR signal. This example is taken from [Don93a].

estimating an unknown signal (function) $f(t)$ from data $y(t)$ where

$$y(t) = f(t) + \epsilon(t) \qquad (1.1)$$

and where the $\epsilon(t)$ are independent normal random variables. In an example taken from [Don93a], figure 1.3 shows a noisy nuclear magnetic resonance (NMR) signal and the WaveShrink estimate of the de-noised NMR signal. WaveShrink is able to remove the noise from the NMR signal while preserving the spike. Traditional noise reduction methods, such as splines, linear filters, or kernel smoothers, would result in some smoothing of the spike.

The WaveShrink algorithm works by taking the wavelet transform of the signal, shrinking the coefficients towards zero, and inverting the transform. The process of shrinking coefficients is much like the process of keeping only important coefficients in data compression algorithms. Like wavelet compression, WaveShrink works well when the original signal has a sparse representation in the wavelet domain.

The original Donoho and Johnstone theory and methodology focused on the Gaussian independent and identically distributed errors setting, as in (1.1). This methodology has been extended to a broader class of models and situations: correlated noise situations [JS94b, NvS95], probability density estimation [JKP92, DJ93, Tri95, VV95], spectral density estimation [Mou92, Gao93b, Gao93a, Mou93, Per93, vSS94, NvS95, WMP95], inverse problems [Don91, MW94, XKS94, Kol94, LM95], classification and discriminant analysis [SC94, BD95], factor analysis [Wic94b], and change point problems [Wan95].

1.3   Example 3: Time-Frequency Analysis

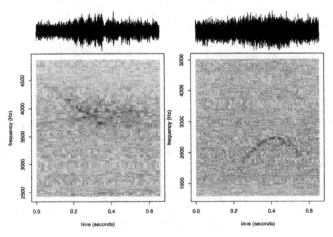

FIGURE 1.4. Time frequency displays for two acoustic signals: a porpoise whistle (left) and an ice squeal (right). The time-frequency display gives an estimate of the local frequency content of a signal at a given time point: the vertical axis is frequency and the horizontal axis is time. These time-frequency displays are based on wavelet packet transforms selected by the Coifman and Wickerhauser "best basis" algorithm.

Fourier analysis is the fundamental tool for understanding the frequency structure of stationary signals. Many signals and time series, however, are non-stationary and their frequency behavior evolves over time. The aim of *time-frequency* analysis is to study the frequency domain properties of non-stationary signals. One approach to time-frequency analysis is to represent the signal as a sum of orthogonal time-frequency waveforms. Two important families of time-frequency waveforms, developed by researchers at Yale University [CMW92], are wavelet packets and cosine packets. These waveforms provide the basis for efficient time-frequency representations for many types of signals. As in data compression and noise removal, the efficiency of the representation is important for the clarity of the analysis.

Figure 1.4 displays a time-frequency analysis of two underwater acoustic signals: a porpoise whistle and an ice squeal. The analysis is based on wavelet packet time-frequency representations selected by the Coifman and Wickerhauser "best basis" algorithm [CW92]. The time-frequency display gives an estimate of the frequency content of a signal locally at a given time point. The vertical axis is frequency and the horizontal axis is time. Both signals contain chirps; the porpoise chirp is close to linear while the ice chirp is clearly quadratic. A harmonic of the chirp for the ice squeal is barely visible. The best basis algorithm automatically adapts the transform to best match the characteristics of the signal. In the above example, it chooses wavelet packets tuned to the different chirps and harmonics observed in the time-frequency displays.

1.4   Example 4: Prototyping Fast Algorithms

For the same reason that wavelets are good for noise removal and for image compression, wavelets are also good as a generic way to speed up and improve existing algorithms. The basic idea is to reduce the dimensionality of the problem by reformulating the problem in a compressed wavelet domain. With wavelets, we have the potential to apply standard statistical or numerical methods, such as matrix multiplication or statistical classification, on a much larger data set than is normally feasible.

A fast wavelet-based classification algorithm was developed by Saito and Coifman [SC94]. This classifier first pre-processes the data with a wavelet transform and then applies a tree network classifier to a few "important" wavelet coefficients. Figure 1.5 illustrates the wavelet classifier applied to noisy signals generated from three signal shapes. This example was originally studied by Breiman, Friedman, Olshen, and Stone [BFOS84]. The classification tree based on the wavelet coefficients is much faster to compute than a tree based on the original samples, and has a lower misclassification rate in this example.

The S+WAVELETS software environment is ideally suited for improving and speeding up statistical algorithms using wavelets. Indeed, a major strength of the S-PLUS environment is the capacity to prototype new algorithms and methods. The interactive and object-

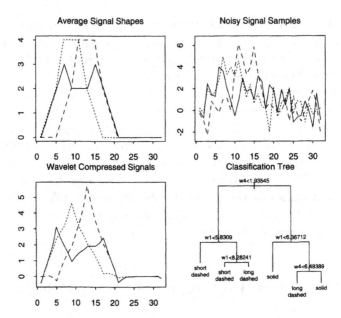

FIGURE 1.5. The problem is to classify noisy signals selected from three signal shapes (upper left) drawn as solid, short dashed, and long dashed lines. Three noisy signals (upper right) are generated from the three shapes. The wavelet transform is applied to each signal and a tree classifier (lower right) is developed based on four wavelet coefficients. This classification tree is much faster to compute than a tree based on the original samples, and has a lower misclassification rate. The reconstructed signals from the four wavelet coefficients (lower left) are much smoother than the original signals.

oriented nature of the language makes it easy to rapidly test new research ideas.

1.5   References for Wavelet Analysis

This book offers a visual approach to exploring wavelet analysis which is accessible to engineers, scientists, and data analysts who have basic undergraduate mathematics training in calculus and linear algebra. We do not cover much of the mathematical and historical development of the field. For a non-technical historical account of wavelets, we recommend Yves Meyer's book *Wavelets: Algorithms and Applications* [Mey93] and Barbara Burke Hubbard's book *The World According to Wavelets* [Bur95].

At a more technical level are several books assuming mathematical training. Of particular note is the book *Ten Lectures on Wavelets* by Ingrid Daubechies [Dau92], which gives a highly readable account of some of the mathematical foundations of wavelet analysis. The book *Adapted Wavelet Analysis: From Theory to Software* by Victor Wickerhauser [Wic94a] gives a good technical discussion of wavelet packet and cosine packet analysis at both the mathematical level and the software engineering level. The book *Wavelets and Subband Coding* by Martin Vetterli and Jelena Kovačević presents wavelets using the language and theory of subband coding. Many other mathematically oriented books have appeared (e.g., [Mey90, AH92, Chu92a, Chu92b, VV93, EGK94, SN96]) and more are soon to appear.

Research activity in wavelet analysis is proceeding at breathtaking pace, and we do not provide a comprehensive survey of all applications of wavelet analysis. An excellent resource for current research activity on wavelet analysis can be found on the Wavelet Digest, an electronic bulletin board for announcements about wavelets conferences, workshops, courses, software packages, and technical papers. It also provides a forum for questions about wavelets, and maintains archives of announcements and articles. The Wavelet Digest can be reached at `http://www.math.scarolina.edu/~wavelets` on the world-wide web.

2

Wavelet Analysis of 1-D Signals

This chapter gives you a tour of wavelet analysis for one-dimensional signals. You will learn how to do the following:

- Create wavelet objects using the `wavelet` function (section 2.1).

- Apply the discrete wavelet transform (DWT) and the inverse DWT to a signal using the `dwt` and `idwt` functions (section 2.2).

- Decompose and approximate a signal into multiresolution layers using the functions `mrd` and `mra` (section 2.3).

- Visualize a time-scale representation of a signal using the function `time.scale.plot` (section 2.4).

- Create and use "biorthogonal" wavelets (section 2.5).

- Use the fast pyramid algorithm to compute the discrete wavelet transform (section 2.6).

2.1 Wavelet Functions and Approximations

Wavelets are fundamental building block functions, analogous to the trigonometric sine and cosine functions. As with a sine or cosine wave, a wavelet function oscillates about zero. However, the oscillations for a wavelet damp down to zero and the function is *localized* in time or space. Hence the name "wavelet."

Create a wavelet with the `wavelet` function:

```
> d4 <- wavelet("d4")
```

This produces the wavelet object **d4**.

The **d4** wavelet, discovered by Ingrid Daubechies [Dau92], is famous for its special mathematical properties and its unusual shape. Plot this wavelet with the `plot` function:

```
> plot(d4)
```

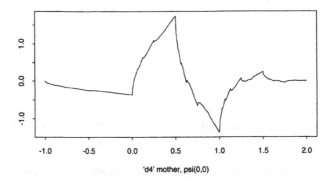

FIGURE 2.1. The famous **d4** wavelet function.

In wavelet analysis, you use linear combinations of wavelet functions to represent signals $f(t)$. These representations are useful in a broad range of applications, such as data compression, signal and image processing, nonparametric statistical estimation, numerical analysis, chemistry, astronomy, and oceanography. Some characteristics which make the wavelet approximation so remarkable and useful are:

- Wavelets are *localized* in time and are good *building block* functions for a variety of signals, including signals with features which change over time and signals which have jumps and other non-smooth features. A traditional Fourier series approximation is not well suited to these types of signals.

- Wavelets separate a signal into *multiresolution* components. The fine and coarse resolution components capture, respectively, the fine and coarse scale features in the signal.

- The wavelet approximation can compact the energy of a signal into a relatively small number of wavelet functions. This *data compression* feature of wavelets is valuable for applications such as nonparametric statistical estimation and classification (see chapter 6).

2.1.1   Father and Mother Wavelets

Wavelets have a gender: there are father wavelets ϕ and mother wavelets ψ. The father wavelet integrates to 1 and the mother wavelet integrates to 0:

$$\int \phi(t)dt = 1 \qquad \int \psi(t)dt = 0.$$

Roughly speaking, the father wavelets are good at representing the smooth and low-frequency parts of a signal and the mother wavelets are good at representing the detail and high-frequency parts of a signal.

Father and mother wavelets pairs come in *families*. There are many different families of wavelets in S+WAVELETS; the orthogonal wavelet families are listed in section 2.1.4. The default wavelet family in S+WAVELETS is the **s8** wavelet.

The **wavelet** function is used to create both mother and father wavelets. By default, **wavelet** produces a mother wavelet. Set the optional argument **mother=F** to produce a father wavelet.

Create a pair of **s8** father or mother wavelets as follows:

```
> s8.f <- wavelet("s8", mother=F)
> s8.m <- wavelet("s8")
```

This produces the father and mother wavelet objects **s8.f** and **s8.m**. Plot these wavelets with the **plot** function:

```
> par(mfrow=c(1,2))
> plot(s8.f)
> plot(s8.m)
```

which gives you the plots in figure 2.2. The **par** command sets up the plotting region to display two plots side-by-side.

Except in some special cases, there is no analytic formula for computing a wavelet function. Instead, wavelet functions are computed using a special two-scale dilation equation (see section 12.5). To learn more about wavelet functions and the dilation equation, refer to the

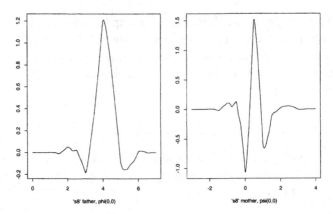

FIGURE 2.2. Father and mother "s8" wavelets.

review article by Gilbert Strang [Str89] and to the books *Ten Lectures in Wavelets* by Ingrid Daubechies [Dau92] and *Introduction to Wavelets* by Charles Chui [Chu92a].

2.1.2   The Wavelet Approximation

The orthogonal wavelet series approximation to a continuous time signal $f(t)$ is given by

$$f(t) \approx \sum_k s_{J,k}\phi_{J,k}(t) + \sum_k d_{J,k}\psi_{J,k}(t) +$$
$$\sum_k d_{J-1,k}\psi_{J-1,k}(t) + \cdots + \sum_k d_{1,k}\psi_{1,k}(t) \quad (2.1)$$

where J is the number of multiresolution components (or *scales*), and k ranges from 1 to the number of coefficients in the specified component. The coefficients $s_{J,k}$, $d_{J,k}$, ..., $d_{1,k}$ are the *wavelet transform coefficients*. The functions $\phi_{J,k}(t)$ and $\psi_{j,k}(t)$ are the approximating wavelet functions.

The functions $\phi_{j,k}(t)$ and $\psi_{j,k}(t)$ are generated from ϕ and ψ through scaling and translation as follows:

$$\phi_{j,k}(t) = 2^{-j/2}\phi(2^{-j}t - k) = 2^{-j/2}\phi\left(\frac{t - 2^j k}{2^j}\right) \quad (2.2)$$

$$\psi_{j,k}(t) = 2^{-j/2}\psi(2^{-j}t - k) = 2^{-j/2}\psi\left(\frac{t - 2^j k}{2^j}\right). \quad (2.3)$$

In S+WAVELETS, the term *wavelet* is used to refer to any one of the functions ψ, ϕ, $\psi_{j,k}$ or $\phi_{j,k}$.

The wavelet coefficients are given approximately by the integrals

$$s_{J,k} \approx \int \phi_{J,k}(t)f(t)dt \tag{2.4}$$

$$d_{j,k} \approx \int \psi_{j,k}(t)f(t)dt, \qquad j = 1, 2, \ldots, J. \tag{2.5}$$

Their magnitude gives a measure of the contribution of the corresponding wavelet function to the approximating sum.

The above wavelet approximation is an orthogonal series approximation since the *basis* functions $\phi_{j,k}(t)$ and $\psi_{j,k}(t)$ are by construction *orthogonal*:

$$
\begin{aligned}
\int \phi_{J,k}(t)\phi_{J,k'}(t)dt &= \delta_{k,k'} \\
\int \psi_{j,k}(t)\phi_{J,k'}(t)dt &= 0 \\
\int \psi_{j,k}(t)\psi_{j',k'}(t)dt &= \delta_{j,j'}\delta_{k,k'}
\end{aligned}
\tag{2.6}
$$

where

$$
\delta_{i,j} = \begin{cases} 1 & \text{if } i = j \\ 0 & \text{if } i \neq j \end{cases}.
$$

There are many other kinds of orthogonal series representations and approximations of a signal $f(t)$. Until quite recently the Fourier series, which uses the *sine* and *cosine* orthogonal basis functions, has been by far the most important such representation in engineering and science applications. Now wavelet series approximations appear to have comparable importance.

2.1.3 Location and Scale Families

The basis functions $\phi_{j,k}(t)$ and $\psi_{j,k}(t)$ in the wavelet approximation (2.1) are *scaled* and *translated* versions of ϕ and ψ, with scale factor 2^j and translation parameter $2^j k$, respectively. The scale factor 2^j is also called the *dilation* factor and the translation parameter $2^j k$ is also called the *location*. In S+WAVELETS, we refer to j as the level index associated with scale 2^j, and k as the shift index associated with translation $2^j k$.

When j gets larger, the scale factor 2^j gets larger, and the functions $\phi_{j,k}(t)$ and $\psi_{j,k}(t)$ get shorter and more spread out, and conversely when j gets smaller. Thus 2^j is a measure of the scale, or width, of the functions $\phi_{j,k}(t)$ and $\psi_{j,k}(t)$. The translation parameter $2^j k$ is matched to the scale parameter 2^j in the sense that as the functions $\phi_{j,k}(t)$ and $\psi_{j,k}(t)$ get wider, their translation steps are correspondingly larger.

To create scaled and translated wavelets $\phi_{j,k}(t)$ and $\psi_{j,k}(t)$, use the optional arguments level and shift. For example, the following commands create and plot two s8 wavelets, one with level 1 and shift 2 (scale 2 and translation 4), and one with level 2 and shift 0 (scale 4 and translation 0):

```
> par(mfrow=c(1,2))
> wave1.2 <- wavelet("s8", level = 1, shift=2)
> wave2.0 <- wavelet("s8", level = 2)
> plot(wave1.2, ylim=c(-.6,.8), xlim=c(-12, 17))
> plot(wave2.0, ylim=c(-.6,.8), xlim=c(-12, 17))
```

This produces the plot of figure 2.3. The arguments xlim and ylim

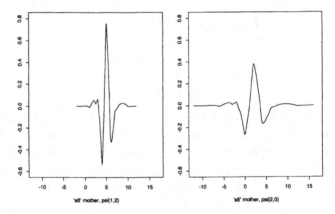

FIGURE 2.3. Scaled and translated "s8" wavelets.

to the plot function ensure that the two plots have the same range for the x and y axes.

2.1.4 Orthogonal Wavelet Families

Only very special pairs or *families* of functions ϕ and ψ result in an orthogonal wavelet series approximation. These special pairs must satisfy exacting mathematical conditions; see [Str89, Dau92, Chu92a] for details. There are numerous types of father and mother wavelets ϕ and ψ in general use, many of which are available in S+WAVELETS.

Type	Wavelet name
Haar	"haar"
Daublets	"d4" "d6" "d8" "d10" "d12" "d14" "d16" "d18" "d20"
Symmlets	"s4" "s6" "s8" "s10" "s12" "s14" "s16" "s18" "s20"
Coiflets	"c6" "c12" "c24" "c30"

TABLE 2.1. Table of orthogonal wavelets available in S+WAVELETS. The number is related to the width and smoothness of the wavelet function. The wavelet "s20" is wider and smoother than the "s8" wavelet.

Table 2.1 lists the four types of orthogonal wavelets currently available in S+WAVELETS.

Haar: The **haar** wavelet is a square wave. It was discovered by the mathematician Haar in 1910 [Haa10], and provided the first known orthogonal wavelet series representation. The **haar** wavelet has *compact support*, that is, it is zero outside a finite interval. It is the only compact orthogonal wavelet which is symmetric. However, unlike the other wavelets, the **haar** wavelet is not continuous.

Daublets: The *daublets* were the first type of *continuous* orthgonal wavelet with compact support. This type of wavelet is named in honor of its discoverer Ingrid Daubechies, who is one of the pioneers in wavelets research.

Symmlets: The *symmlets* also have compact support, and were also constructed by Daubechies [Dau92]. While the *daublets* are quite asymmetric, the *symmlets* were constructed to be as nearly symmetric (least asymmetric) as possible. The default wavelet in S+WAVELETS is the **s8** *symmlet*.

Coiflets: The *coiflets* were constructed by Daubechies to be nearly symmetric and also have additional properties thought to be desirable (vanishing moments for both ϕ and ψ). Daubechies used the name *coiflets* in honor of Ronald Coifman, another important contributor to the theory and application of wavelet analysis.

The first letter of the wavelet indicates the name: **d** for daublet, **s** for symmlet, and **c** for coiflet. The number of the wavelet indicates its width and smoothness. Wavelets with large numbers, such as **d20** or **c30**, are relatively wide and smooth. Wavelets with small numbers, such as **d4** or **c6**, are narrower and less smooth. See section 5.1.2, Selecting a Wavelet, to find out more about what the number means and how it relates to width, smoothness, and other properties.

Let us look at some different orthogonal wavelets. The following commands give the plots of mother wavelets **"haar"**, **"d12"**, **"s12"**, and **"c12"** shown in figure 2.4:

```
> par(mfrow=c(2,2))
> plot(wavelet("haar"))
> plot(wavelet("d12"))
> plot(wavelet("s12"))
> plot(wavelet("c12"))
```

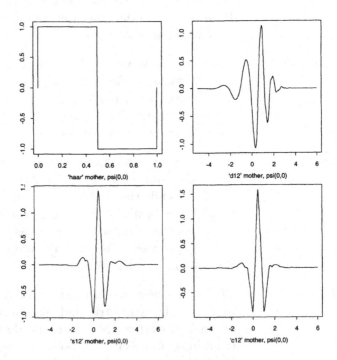

FIGURE 2.4. Four different orthogonal mother wavelets **"haar"**, **"d12"**, **"s12"**, and **"c12"**

2.2   The Discrete Wavelet Transform

The discrete wavelet transform (DWT) calculates the coefficients of the wavelet series approximation (2.1) for a *discrete* signal $f_1, \ldots, f_n$ of finite extent. The DWT maps the vector $\mathbf{f} = (f_1, f_2, \ldots, f_n)'$ to a vector of n wavelet coefficients $\mathbf{w} = (w_1, w_2, \ldots, w_n)'$.

The vector $\mathbf{w}$ contains the coefficients $s_{J,k}$ and $d_{j,k}, j = 1, 2, \ldots, J$ of the wavelet series approximation (2.1). The $s_{J,k}$ are called the *smooth* coefficients and the $d_{j,k}$ are called the *detail* coefficients. Roughly speaking, the smooth coefficients $s_{J,k}$ are thought to represent the underlying smooth behavior of the data at the coarse scale 2^J. The detail coefficients $d_{J,k}$ provide the coarse scale deviations from the smooth behavior. The detail coefficients $d_{J-1,k}, d_{J-2,k}, \ldots, d_{1,k}$ represent progressively finer scale deviations from the smooth behavior.

The DWT is mathematically equivalent to multiplication by an orthogonal matrix $\mathbf{W}$:

$$\mathbf{w} = \mathbf{Wf}. \tag{2.7}$$

To compute the DWT, you don't actually perform the matrix multiplication (2.7). Instead, you use a fast "pyramid" algorithm with complexity $O(n)$, which is even faster than the fast Fourier transform (FFT)! To learn about the pyramid algorithm, turn to section 2.6.

Note: You can think of the discrete signal $f_1, f_2, \ldots, f_n$ as being obtained by sampling a continuous time signal $f(t)$ at a sampling interval of Δ, so $f_i = f(i\Delta)$ for $i = 1, 2, \ldots, n$. The discrete wavelet transform then leads to the wavelet approximation of (2.1). The resolution of this approximation is roughly equal to the sampling interval. Researchers are investigating different ways to initialize and compute the discrete wavelet transform from continuous signals. These methods may be more appropriate for some problems. To find out more about this aspect of wavelet analysis, refer to [She92, Don92, SP92].

2.2.1   Computing the Discrete Wavelet Transform

To illustrate the discrete wavelet transform, we create and plot the doppler signal

```
> doppler <- make.signal("doppler",n=1024)
> plot(doppler, type="l")
```

which is displayed in figure 2.5. The `doppler` signal is a sinusoid with a changing amplitude and decreasing frequency defined by

$$f(t) = \text{doppler}(t) = \sqrt{t(1-t)} \sin\left(\frac{2.1\pi}{t+0.05}\right) \qquad 0 \le t \le 1.$$

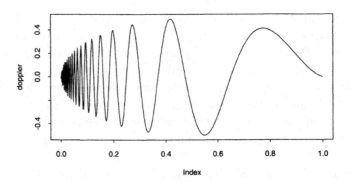

FIGURE 2.5. The `doppler` signal, which is a sinusoid with a changing amplitude and decreasing frequency.

The S+WAVELETS function `dwt` computes the discrete wavelet transform. The discrete wavelet transform of the `doppler` signal is computed and plotted as follows:

```
> doppler.dwt <- dwt(doppler)
> plot(doppler.dwt)
```

Figure 2.6 displays the resulting plot. The original signal is plotted in the top row. The wavelet coefficients are plotted in the remaining rows, going from the fine scale $d_{1,k}$ coefficients in the second row to the coarse scale coefficients $d_{J,k}$ and $s_{J,k}$ in the bottom two rows. The coefficients are plotted as vertical lines extending from zero.

In figure 2.6, the original signal is actually recomputed from the wavelet coefficients by means of the inverse discrete wavelet transform (IDWT), as decsribed in section 2.2.3. This is why the label `idwt` is used in this plot.

In the case that n is divisible by 2^J, there are $n/2$ coefficients $d_{1,k}$ at the finest scale $2^1 = 2$. At the next finest scale $2^2 = 4$, there are $n/4$ coefficients $d_{2,k}$. Likewise, at the coarsest scale, there are $n/2^J$ coefficients each for $d_{J,k}$ and $s_{J,k}$. Summing these up we have a total of n coefficients:

$$n = n/2 + n/4 + \cdots + n/2^{J-1} + n/2^J + n/2^J. \qquad (2.8)$$

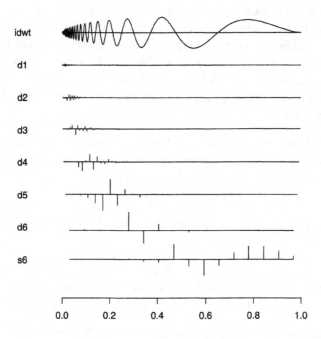

FIGURE 2.6. DWT of the `doppler` signal using default `s8` wavelet.

The number of coefficients at a scale is related to the width of the wavelet function. At scale 2, the translation steps are $2k$, and so $n/2$ terms are required in order for the functions $\psi_{1,k}(t)$ to *cover* the interval $1 \leq t \leq n$. By similar reasoning, a summation involving $\psi_{j,k}(t)$ requires only $n/2^j$ terms, and the summation involving $\phi_{J,k}(t)$ requires only $n/2^J$ terms.

Note: In (2.8), it is assumed that n is divisible by 2^J. If this is not the case, then there are roughly (but not exactly) $n/2^j$ wavelet coefficients at scale 2^j. Nonetheless, the total number of wavelet coefficients is still n.

The coefficients are plotted at approximately the position of the corresponding wavelet function. When n is a power of 2, the plotting position of the (j, k) coefficient is $2^j \left(k + \frac{1}{2} \right)$. In the **haar** wavelet case, this corresponds to the middle of the associated wavelet function.

By default, the detail coefficients $d_{J,k}, d_{J-1,k}, \ldots, d_{1,k}$ are plotted on the same vertical scale. You can measure the relative importance of the detail coefficients by comparing the magnitudes of the coefficients in the plot. The original signal and wavelet coefficients $s_{J,k}$ are plotted on different vertical scales, because the values of the original signal and the values of the $s_{J,k}$ coefficients often have a much larger dynamic range than the detail coefficients.

Observe the following properties of the DWT coefficients for the doppler signal:

1. Typically, the wavelet coefficients at coarse scales are larger than the wavelet coefficients at fine scales. This is a consequence of the smoothness of the doppler signal.

2. The smooth coefficients $s_{6,k}$ correspond to the smooth at scale 2^6, mainly capturing the low frequency oscillations in the latter portion of the signal.

3. The detail coefficients $d_{6,k}$, $d_{5,k}$, ..., $d_{1,k}$ represent progressively finer "corrections" to the smooth trend, capturing the higher frequency oscillations in the beginning of the signal.

4. The coefficients are sparse in the sense that many coefficients are very small or nearly zero.

2.2.2   DWT Objects

The dwt function produces a DWT object which has class dwt. One gets a description of the DWT object doppler.dwt by typing the object name:

```
> doppler.dwt
Discrete Wavelet Transform of doppler
Wavelet: s8
Length of series: 1024
Number of levels: 6
Boundary correction rule: periodic
Crystals: s6 d6 d5 d4 d3 d2 d1
```

The default discrete wavelet transform uses the s8 wavelet, computes 6 levels (scales), and uses the "periodic" boundary correction rule. Turn to section 5.1 to see how to change these defaults.

The doppler.dwt object, like most DWT objects, is a vector of wavelet coefficients. The coefficients are ordered from coarse scales to fine scales in the vector $\mathbf{w}$. In the case where n is divisible by 2^J,

$$
\mathbf{w} = \begin{pmatrix} \mathbf{s}_J \\ \mathbf{d}_J \\ \mathbf{d}_{J-1} \\ \vdots \\ \mathbf{d}_1 \end{pmatrix} \tag{2.9}
$$

where

$$
\begin{aligned}
\mathbf{s}_J &= (s_{J,1}, s_{J,2}, \cdots, s_{J,n/2^J})' \\
\mathbf{d}_J &= (d_{J,1}, d_{J,2}, \cdots, d_{J,n/2^J})' \\
\mathbf{d}_{J-1} &= (d_{J-1,1}, d_{J-1,2}, \cdots, d_{J-1,n/2^{J-1}})' \\
\vdots &= \quad \vdots \\
\mathbf{d}_1 &= (d_{1,1}, d_{1,2}, \cdots, d_{1,n/2})'.
\end{aligned} \tag{2.10}
$$

Each of the sets of coefficients $\mathbf{s}_6, \mathbf{d}_6, \ldots, \mathbf{d}_1$ is called a *crystal*. The term crystal is used because the wavelet coefficients in a crystal correspond to a set of translated wavelet functions arranged on a regular lattice. A crystal is commonly referred to as a *subband*, a term which comes from the close link between wavelets and "subband filtering" (see chapter 5 of [Dau92]).

To extract a crystal, use the subscript operator [[:

```
> doppler.dwt[["s6"]]
      s6(1)         s6(2)        s6(3)      s6(4)       s6(5)
 0.007765673 -0.001330175 -0.07550149 0.110322 0.1329815
      s6(6)         s6(7)     s6(8)       s6(9)      s6(10)
  -0.466046 -0.6727822 3.648829 -1.715457 -4.011206
      s6(11)      s6(12)    s6(13)   s6(14)    s6(15)      s6(16)
  -1.536702 1.623148 3.211735 3.10738 2.040245 0.7879406
```

Note that some crystals are very long vectors, and you probably don't want to print them. For example, the $\mathbf{d}_1$ crystal for the `doppler.dwt` object is a vector of length 512.

The `dictionary` attribute of a DWT object retains information about the transform including the type of wavelet, boundary rule, number of resolution levels, and wavelet filters. DWT objects have other attributes, such as the names and lengths of the crystals. In chapter 5, you can find out more about how to subscript and manipulate DWT objects.

2.2.3   The Inverse Discrete Wavelet Transform

You can recover the original signal vector **f** by applying the inverse discrete wavelet transform (IDWT). The IDWT is implemented in S+WAVELETS with the function idwt:

```
> doppler.recon <- idwt(doppler.dwt)
```

Due to round-off error, the reconstructed signal doppler.recon is not identically equal to the original doppler signal. To assess the round-off error of the reconstruction $\widehat{f}_i$, look at the L^2 relative error

$$\text{relative error} = \frac{\left(\sum_{i=1}^n (f_i - \widehat{f}_i)^2\right)^{1/2}}{\left(\sum_{i=1}^n f_i^2\right)^{1/2}}. \tag{2.11}$$

Compute the relative error for doppler.recon using the function **vecnorm**:

```
> vecnorm(doppler - doppler.recon)/vecnorm(doppler)
[1] 2.462808e-12
```

Note: You can also compute the inverse transform using the generic function reconstruct, which also performs reconstruction computations on other kinds of S+WAVELETS objects.

2.2.4   The Wavelet Approximation Using the Largest Coefficients

The wavelet approximation is useful because wavelet functions are very good building blocks for a variety of different signals and functions and, in general, you can reconstruct the original signal from a small number of coefficients. This section illustrates these qualities using the wavelet approximation for the ramp signal.

Create a ramp signal of length $n = 256$ and take its DWT:

```
> ramp <- make.signal("ramp", n=256)
> ramp.dwt <- dwt(ramp, wavelet="d4")
> plot(ramp.dwt)
```

The plot is shown in figure 2.7. The DWT of the ramp signal is very sparse in the sense that almost all of the coefficients are zero. The DWT compacts the energy of the ramp signal into just a few coefficients.

Extract the largest 16 coefficients using the function top.atoms:

```
> ramp.top <- top.atoms(ramp.dwt, n.atoms=16)
```

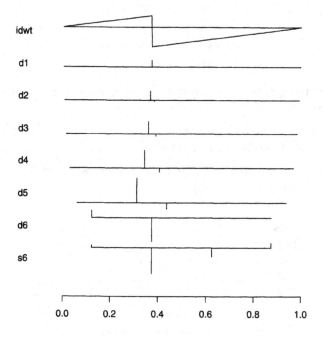

FIGURE 2.7. DWT of the `ramp` signal.

The object `ramp.top` is a "wavelet molecule." It is called a molecule since the coefficients correspond to a list of atomic wavelet functions not organized on a lattice. You can learn more about molecules in section 10.3.

We now reconstruct the ramp signal from the 16 coefficients held in `ramp.top`:

```
> ramp.recon <- reconstruct(ramp.top)
> vecnorm(ramp.recon - ramp)/vecnorm(ramp)
[1] 3.35973e-16
```

We get almost perfect reconstruction of the ramp function of length $n = 256$ from just 16 wavelets! The small error is due to numerical round-off error.

In terms of the wavelet approximation, we are able to decompose the ramp signal as a sum of 16 individual scaled wavelet functions:

$$f(t) = S_{6,4}(t) + S_{6,3}(t) + \cdots + D_{1,65}(t)$$

where the scaled wavelet functions $S_{J,k}$ and $D_{j,k}$ are defined as follows:

$$S_{J,k}(t) = s_{J,k}\phi_{J,k}(t) \qquad (2.12)$$
$$D_{j,k}(t) = d_{j,k}\psi_{j,k}(t). \qquad (2.13)$$

To visualize this decomposition, apply the **decompose** function to produce figure 2.8:

```
> ramp.decomp <- decompose(ramp.top)
> plot(ramp.decomp, n.top=16)
```

This plots the top 16 wavelets in order from most significant $(S_{6,4}(t))$ to least significant $(D_{1,65}(t))$.

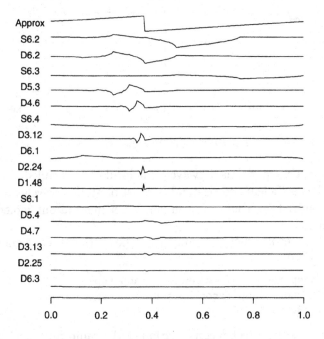

FIGURE 2.8. Decomposition of the **ramp** signal into the top 16 wavelet functions.

The **decompose** function can be used to produce other kinds of decompositions. Read the next section to find out about decompositions which arise from *multiresolution analysis*.

2.3   Multiresolution Analysis

The functions

$$S_J(t) = \sum_k s_{J,k}\phi_{J,k}(t) \tag{2.14}$$

and

$$D_j(t) = \sum_k d_{j,k}\psi_{j,k}(t) \tag{2.15}$$

are called the *smooth* signal and the *detail* signals respectively. The orthogonal wavelet series approximation to a continuous signal $f(t)$ is expressed in terms of these signals:

$$f(t) \approx S_J(t) + D_J(t) + D_{J-1}(t) + \cdots + D_1(t).$$

The terms in this approximating sum constitute a *decomposition* of the signal into orthogonal signal *components* $S_J(t)$, $D_J(t)$, $D_{J-1}(t)$, ..., $D_1(t)$ at different scales. Because the terms at different scales represent components of the signal $f(t)$ at different *resolutions*, the approximation is called a *multiresolution decomposition* (MRD).

The theory of multiresolution analysis was first investigated by Stéphane Mallat [Mal89b, Mal89a] and Yves Meyer [Mey86]. To find out more about multiresolution analysis, refer to one of these articles or to chapter 5 of the book by Daubechies [Dau92].

2.3.1   Computing a Multiresolution Decomposition

Use **mrd** to obtain a multiresolution decomposition for the **doppler** signal as shown in figure 2.9:

```
> doppler.mrd <- mrd(doppler)
> plot(doppler.mrd)
```

The fine scale features (the high frequency oscillations at the beginning of the signal) are captured mainly by the fine scale detail components D_1 and D_2. The coarse scale components D_6 and S_6 correspond to lower frequency oscillations towards the end of the series.

Print the **doppler.mrd** object:

```
> doppler.mrd
Wavelet Decomposition of doppler
Wavelet: s8
Length of series: 1024
```

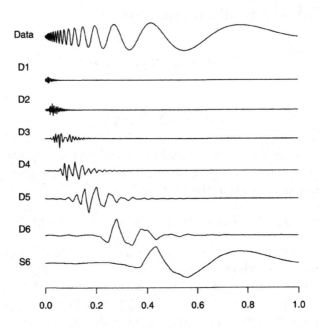

FIGURE 2.9. Multiresolution decomposition of the doppler signal.

```
Number of levels: 6
Boundary correction rule: periodic
Signal Components: D1 D2 D3 D4 D5 D6 S6
```

As a mnemonic, uppercase letters are used for the names of signal components while lowercase letters are use for the names of coefficient crystals. For example, d1 refers to the coefficients $d_{1,k}$ and D1 refers to signal $D_1(t)$.

The doppler.mrd object is a matrix with $n = 1024$ rows and $m = 7$ columns which has class decompose. The columns correspond to the signals $S_6(t), D_6(t), D_5(t), \ldots, D_1(t)$ defined by (2.14) and (2.15). You can extract the signal components using the [[operator. For example, the expression doppler.mrd[["D3"]] extracts the $D_3(t)$ signal component.

Note: Instead of using mrd, you can create doppler.mrd by first creating a dwt object and then applying the decompose function:

```
> doppler.dwt <- dwt(doppler)
> doppler.mrd <- decompose(doppler.dwt)
```

2.3.2   Multiresolution Approximation

The coarsest scale signal $S_J(t)$ gives a coarse scale smooth approximation to your signal $f(t)$. Adding the detail signal $D_J(t)$ yields $S_{J-1}(t)$, a scale 2^{J-1} approximation to the signal. The $S_{J-1}(t)$ approximation is a refinement of the $S_J(t)$ approximation. Similarly, you can refine further to obtain the scale 2^{j-1} approximations

$$S_{j-1}(t) = S_J(t) + D_J(t) + \cdots + D_j(t).$$

The collection S_J, S_{J-1}, ..., S_1 provides a set of *multiresolution approximations* of $f(t)$. The approximations range from coarse to fine scales 2^J, 2^{J-1}, ... , 2. The **mra** functon computes a set of multiresolution approximations for the **doppler** signal:

```
> doppler.mra <- mra(doppler)
> plot(doppler.mra)
```

The multiresolution approximation plot is given in figure 2.10.

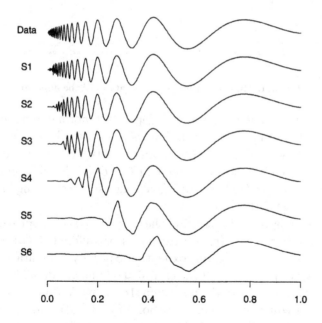

FIGURE 2.10. Multiresolution approximations of the **doppler** signal.

2.3.3   Zooming Feature of Multiresolution Analysis

Multiresolution analysis gives you the framework in which to zoom in
and out to obtain different views of a signal. To illustrate the zooming
concept, add some random noise to the doppler signal and perform
a multiresolution analysis of the noisy signal to obtain figure 2.11:

```
> par(mfrow=c(1,2))
> noisy.doppler <- doppler + rnorm(1024)/15
> plot(mrd(noisy.doppler, n.levels=4))
> plot(mra(noisy.doppler, n.levels=4))
```

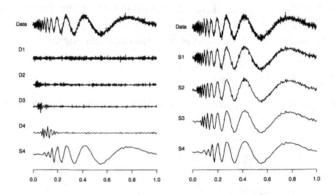

FIGURE 2.11. Multiresolution approximations of the doppler signal.

The rnorm function generates independent normal random noise.
The number of multiresolution levels is restricted by n.levels=4.

   The fine scale detail $D_1(t)$ is almost entirely dominated by the
noise. The noise is still dominant in the $D_2(t)$ detail signal, is less
prominent in the $D_3(t)$ detail signal, and is almost negligible in the
coarsest scale detail $D_4(t)$.

   By zooming in and out using the multiresolution approximations
$S_1(t)$, $S_2(t)$, $S_3(t)$, $S_4(t)$, we can focus on different features of the
signal. The coarse scale approximation $S_4(t)$ gives a "faraway view"
of the signal: you can see the overall shape but some detail is lost. The
$S_4(t)$ approximation of the noisy doppler signal is almost identical to
the lower frequency portion of the noise-free doppler signal; compare
with figure 2.10. The fine scale approximations give a "close-in view"
of the signal, revealing more detail but also more noise.

2.4   Time-Scale View of Wavelets

The wavelet approximation (2.1) separates a signal into different scales or multiresolution levels. *Time-scale analysis* is about studying how the scale representation of the signal changes over time. Time-scale analysis is closely related to *time-frequency analysis*, and is a widely used technique in signal processing. To learn more about time-scale analysis, refer to the review article by Rioul and Vetterli [RV91].

In time-scale analysis, the signal is represented in the "time-scale plane," in which each wavelet coefficient occupies a rectangle. You can visualize the time-scale plane by plotting the modulus of the co-efficients. The time-scale plane is plotted in S+WAVELETS with the function `time.scale.plot`. This is a discrete version of a scalogram [RV91]. To illustrate the time-scale plot, we create and plot a `dirac` signal:

```
> dirac <- make.signal("dirac",n=1024)
> plot(dirac, type="l")
```

The result is shown in figure 2.12.   The `dirac` signal consists of a

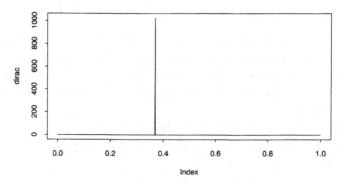

FIGURE 2.12. A `dirac` signal of length $n = 1024$.

single non-zero value at $t \approx 0.37$. Plot the time-scale representation for the `dirac` signal as follows:

```
> time.scale.plot(dwt(dirac))
```

The result is shown in figure 2.13.   The x axis represents time, while the y axis represents 1/scale and ranges from 0 (coarsest scale) to 1 (finest scale).

In the time-scale plane, each wavelet coefficient occupies a box having a constant area. The fine scale wavelet coefficients $d_{1,k}$ oc-

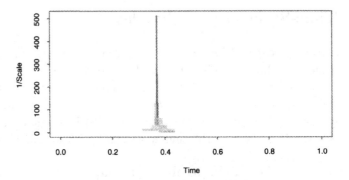

FIGURE 2.13. Time-scale plot giving for a `dirac` signal.

cupy tall thin boxes while the coarse scale wavelet coefficients $d_{J,k}$ occupy flat wide boxes. The coordinates of the boxes for the detail coefficients $d_{j,k}$ are

$$(x_1, x_2, y_1, y_2) = ((k-1)2^{-j}, k2^{-j}, 2^{-j}, 2^{-(j-1)}).$$

The coordinates of the boxes for the smooth coefficients are $s_{J,k}$

$$(x_1, x_2, y_1, y_2) = ((k-1)2^{-J}, k2^{-J}, 0, 2^{-J}).$$

The `dirac` signal is well localized at the fine scales (top of the plot), but is poorly localized at the coarse scales (bottom of the plot). This is because at the finest scale, there are $1024/2 = 512$ coefficients d_1, while at the coarsest scale, there are $1024/2^6 = 16$ coefficients each for d_J and s_J.

Next we create and plot the `jumpsine` signal:

```
> jumpsine <- make.signal("jumpsine",n=1024)
> plot(jumpsine,type="l")
```

The `jumpsine` signal, shown in figure 2.14, is a sine curve with two discontinuous jumps. The time-scale representation is obtained by

```
> time.scale.plot(dwt(jumpsine))
```

which results in the plot shown in figure 2.15. In this plot, you see two things:

1. The jumps show up as fine scale features (the dark vertical regions in the upper part of the plot).

2. The smooth part of the signal shows up as a coarse scale feature (the dark horizontal strip in the bottom of the plot).

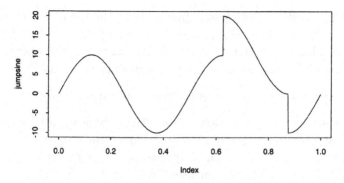

FIGURE 2.14. The **jumpsine** signal.

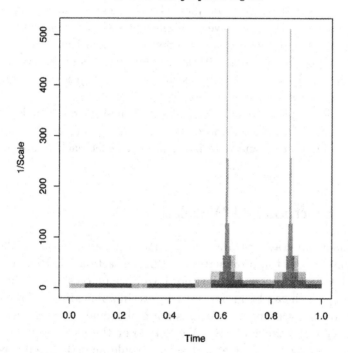

FIGURE 2.15. Time-scale plot for the **jumpsine** signal.

In general, the time-scale plot is good for analyzing signals which exhibit different scale characteristics over time, as with the **jumpsine** example.

Like the **dirac** signal, coarse scale features are poorly localized in time. However, coarse scale features are well localized in scale (frequency): the smooth part of the **jumpsine** signal is represented by a narrow strip. Conversely, fine scale features are well localized in time, but poorly localized in scale (frequency).

The tradeoff between time and scale localization is best understood in terms of the fundamental signal uncertainty relationship:

$$\text{time bandwidth} \times \text{inverse scale bandwidth} \geq \text{constant}.$$

This relationship is a version of the "Heisenberg uncertainty principle." It says that a signal cannot be simultaneously concentrated in arbitrarily small time and reciprocal scale (frequency) regions. If a signal is very highly concentrated in a coarse scale region, then it must be spread out in time.

In terms of the time-scale plot, the Heisenberg uncertainty principle means that each wavelet coefficient occupies a box having a constant area. These boxes are sometimes called Heisenberg uncertainty boxes. For a signal with a sampling interval of Δ_t, the area of each box is $1/\Delta_t$ (for the time-scale plot of the jumpsine signal in figure 2.15, the area of each box is $1/n$).

Time-scale plots are closely related to time-frequency analysis and spectograms. To find out more about this aspect of wavelet analysis, turn to chapters 7 and 8 which discuss wavelet packets and cosine packets.

2.5 Biorthogonal Wavelets

Biorthogonal wavelets are an important generalization of the orthogonal wavelet approximation of (2.1) of section 2.1. Biorthogonal wavelets are symmetric and do not introduce phase shifts in the coefficients, which can be important for certain applications. However, a biorthogonal wavelet transform is not orthogonal. Table 2.2 lists the biorthogonal wavelets in S+WAVELETSFor the most part, you can use a biorthogonal wavelet just as you would an orthogonal wavelet.

This section gives some background on biorthogonal wavelets, including why they are called "biorthogonal." Biorthogonal wavelets with finite support were first introduced by Cohen, Daubechies, and Feauveau [CDF92]. To learn more about the theory of biorthogonal wavelets, refer to [CDF92], chapter 8 of the book by Daubechies [Dau92], or the review article by Jawerth and Sweldens [JS94a].

Biorthogonal wavelet analysis starts with four basic function types: ϕ, ψ, $\tilde{\phi}$, and $\tilde{\psi}$. The functions ϕ and ψ are the usual father and mother

Type	Wavelet name
B-spline	`"bs1.1"` `"bs1.3"` `"bs1.5"`
	`"bs2.2"` `"bs2.4"` `"bs2.6"` `"bs2.8"`
	`"bs3.1"` `"bs3.3"` `"bs3.5"` `"bs3.7"` `"bs3.9"`
V-spline	`"vs1"` `"vs2"` `"vs3"`

TABLE 2.2. Table of biorthogonal wavelets available in S+WAVELETS.

wavelets. The functions $\widetilde{\phi}$ and $\widetilde{\psi}$ are called the *dual* wavelets.

The father and mother wavelets ϕ and ψ play the same role as before in computing the wavelet coefficients—see (2.4) and (2.5). However, the biorthogonal wavelet approximation is expressed in terms of the dual wavelet functions:

$$
f(t) \approx \sum_k s_{J,k}\widetilde{\phi}_{J,k}(t) + \sum_k d_{J,k}\widetilde{\psi}_{J,k}(t) +
$$
$$
\sum_k d_{J-1,k}\widetilde{\psi}_{J-1,k}(t) + \cdots + \sum_k d_{1,k}\widetilde{\psi}_{1,k}(t). \quad (2.16)
$$

Hence, ϕ and ψ are used to "analyze" the signal while $\widetilde{\phi}$ and $\widetilde{\psi}$ are used to "synthesize" the signal. Note that you obtain the orthogonal wavelet approximation (2.1) by setting $\widetilde{\phi} = \phi$ and $\widetilde{\psi} = \psi$.

S+WAVELETS makes available two types of biorthogonal wavelets (see [CDF92, Dau92]):

B-spline: These are based on the simple polynomial spline functions. The wavelets ϕ and ψ can have degree 0 (the **haar** father wavelet), degree 1 (a triangle function), or degree 2 (a quadratic bump function). Several pairs of dual wavelets $\widetilde{\phi}$ and $\widetilde{\psi}$ are available for each pair of wavelets ϕ and ψ. The dual wavelets have different support length. The name of the *b-spline* wavelet has two numbers. The first number indicates the degree of the polynomial for the wavelets (the wavelet for `"bs2.6"` has degree 1). The second number indicates the length of the support of the dual wavelet (the dual wavelet for `"bs2.8"` has wider support than the dual wavelet for `"bs2.6"`). See Chapter 8 of [Dau92] for details.

V-spline: These are "variations" on the *b-spline* de-
                                signed to achieve near-orthogonality and sup-
                                port width which is nearly the same for father
                                and mother wavelets. There are three differ-
                                ent *v-splines*. The numbers of the *v-splines*
                                have no particular meaning.

To create biorthogonal wavelets, use the **wavelet** function. The
dual wavelets are created by setting the optional argument **dual=T**.
Plot the **"bs3.7"** father and mother wavelets and their dual wavelet
functions as follows:

```
> par(mfrow=c(2,2))
> plot(wavelet("bs3.7", mother=F))
> plot(wavelet("bs3.7"))
> plot(wavelet("bs3.7", mother=F, dual=T))
> plot(wavelet("bs3.7", dual=T))
```

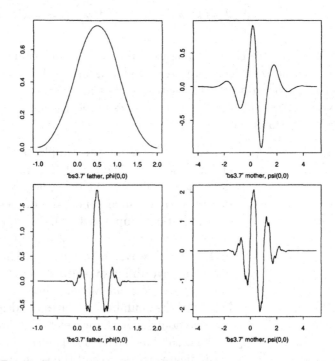

FIGURE 2.16. Father and mother biorthogonal wavelets **"bs3.7"** are plotted
on top left and right respectively. The corresponding dual wavelet functions are
plotted below.

In general, biorthogonal wavelets are not orthogonal and do not satisfy the orthogonality relationships in (2.6). However, they do satisfy a *biorthogonal* relationship

$$\int \phi_{J,k}(t)\widetilde{\phi}_{J,k'}(t)dt = \delta_{k,k'}$$

$$\int \psi_{j,k}(t)\widetilde{\phi}_{J,k'}(t)dt = 0 \qquad (2.17)$$

$$\int \psi_{j,k}(t)\widetilde{\psi}_{j',k'}(t)dt = \delta_{j,j'}\delta_{k,k'}.$$

This relationship is important for the reconstruction formula for biorthogonal wavelets.

Note: The father and mother wavelets and their duals can be reversed. You can use ϕ and ψ in the wavelet approximation and the dual wavelets $\widetilde{\phi}$ and $\widetilde{\psi}$ to compute the wavelet coefficients. In other words, you can use the original wavelet approximation (2.1) but compute the wavelet coefficients by (2.4) and (2.5) with ϕ and ψ replaced by $\widetilde{\phi}$ and $\widetilde{\psi}$.

In the functions dwt, mrd, and mra, set the argument dual=TRUE to use the dual wavelets $\widetilde{\phi}$ and $\widetilde{\psi}$ to analyze the signal (the default is dual=FALSE).

2.6   Pyramid Algorithm

The discrete wavelet transform (DWT) computed by dwt and the inverse discrete wavelet transform (IDWT) computed by the idwt function use Mallat's [Mal89b] remarkably fast pyramid algorithms. These algorithms involve low-pass and high-pass filters, along with a *down-sampling* (decimation) or *up-sampling* (zero-padding) operator.

Mallat's wavelet pyramid algorithm has its roots in the pyramid algorithm of Burt and Adelson [BA83]. Chapter 3 of [Mey93] presents the pyramid algorithm from a historical perspective.

2.6.1   Forward Algorithm to Compute the DWT

The DWT pyramid algorithm is represented by figure 2.17. The basic components in each stage of the pyramid are two *analysis* filters—a low-pass filter L and a high-pass filter H—and a *decimation-by-two* operation.

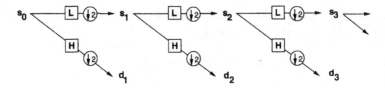

FIGURE 2.17. DWT pyramid algorithm.

The decimation operation, indicated by a downward pointing arrow with the number 2, consists of deleting every other value of the filter outputs.

The input $\mathbf{s}_0 = (s_{0,1}, s_{0,2}, \ldots, s_{0,n})'$ to the pyramid algorithm consists of the values of the discrete signal:

$$s_{0,i} = f_i \qquad i = 1, 2, \ldots, n.$$

Let $n_j = n/2^j$. The output of the algorithm is the set of DWT detail coefficients

$$\mathbf{d}_j = (d_{j,1}, d_{j,2}, \ldots, d_{j,n_j})'$$

at levels $j = 1, 2, \ldots, J$, which correspond to scales $2, 4, 8, \ldots, 2^J$, along with the DWT smooth coefficients

$$\mathbf{s}_J = (s_{J,1}, s_{J,2}, \ldots, s_{J,n_J})'.$$

Since the low-pass and high-pass filters L and H have a fixed number of coefficients (typically quite short), the algorithm has complexity $O(n)$.

2.6.2   Backward Algorithm to Compute the IDWT

The inverse discrete wavelet transform (IDWT) algorithm, represented by figure 2.18, inverts the DWT pyramid algorithm in a straightforward manner.   The basic components in each stage of

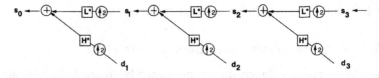

FIGURE 2.18. Reconstruction algorithm for the discrete wavelet transform.

the reconstruction are the *synthesis* low-pass and high-pass filters

L^* and H^* and an *up-sample-by-two* operation. The up-sample operation, indicated by an upward pointing arrow with the number 2, consists of inserting zeros between every other value of the filter inputs. The reconstruction algorithm also has complexity $O(n)$.

Turn to chapter 12 to learn more about wavelet algorithms, including how to create and use objects such as wavelet filters, transfer functions, and orthogonal wavelet matrix transforms $\mathbf{W}$.

3
Wavelet Analysis of Images

In this chapter, you will learn about wavelet analysis for images and matrices, including how to do the following:

- Create a two-dimensional (2-D) wavelet object using the function `wavelet.2d` (section 3.1).

- Apply the 2-D discrete wavelet transform (DWT) and the inverse 2-D DWT to an image using the `dwt.2d` and `reconstruct` functions (section 3.2).

- Decompose and approximate an image into multiresolution layers (section 3.3).

For additional background on 2-D wavelet representations, refer to either the article by Mallat [Mal89a] or chapter 10 of the book by Daubechies [Dau92].

3.1 2-D Wavelet Functions

Two dimensional (2-D) wavelets are used in applications involving images, matrices, and other 2-D data. The properties which make wavelets attractive for analysis of 1-D functions hold for 2-D functions as well. In particular, wavelets have been found to be very effective for image coding and data compression, since relatively few coefficients are needed to represent the image [Mal89b].

Create and plot a 2-D wavelet using the `wavelet.2d` function:

```
> plot(wavelet.2d(wavelet="s8"))
```

This produces the 2-D wavelet pictured in figure 3.1.

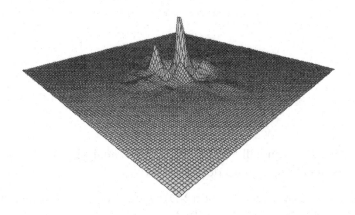

FIGURE 3.1. Plot of a 2-D s8 wavelet function.

3.1.1 2-D Wavelet Families

You can construct 2-D wavelets by taking the tensor product of a *horizontal* 1-D wavelet and a *vertical* 1-D wavelet. This leads to four different types of 2-D wavelets:

$$
\begin{aligned}
\Phi(x,y)   &= \phi_h(x) \times \phi_v(y) &&= \text{horiz. father} &\times& \text{ vert. father} \\
\Psi^v(x,y) &= \psi_h(x) \times \phi_v(y) &&= \text{horiz. mother} &\times& \text{ vert. father} \\
\Psi^h(x,y) &= \phi_h(x) \times \psi_v(y) &&= \text{horiz. father} &\times& \text{ vert. mother} \\
\Psi^d(x,y) &= \psi_h(x) \times \psi_v(y) &&= \text{horiz. mother} &\times& \text{ vert. mother}
\end{aligned}
$$

The 2-D wavelet family has one father wavelet function and three mother wavelet functions. As with 1-D wavelets, the father wavelets are good at representing the smooth and the mother wavelets are good at representing the detail. Roughly speaking, Φ captures the smooth part, Ψ^v captures the vertical detail, Ψ^h captures the horizontal detail, and Ψ^d captures the diagonal detail.

Create a family of 2-D **haar** wavelets as follows:

```
> Phi <- wavelet.2d("haar", mother=c(F, F))
> Psi.v <- wavelet.2d("haar", mother=c(T, F))
> Psi.h <- wavelet.2d("haar", mother=c(F, T))
> Psi.d <- wavelet.2d("haar", mother=c(T, T))
```

Plot the family as follows to produce figure 3.2.

```
> eye <- c(-6,-8,5)
> plot(Phi, axes=F, box=F, eye=eye, J=4)
> plot(Psi.v, axes=F, box=F, eye=eye, J=4)
> plot(Psi.h, axes=F, box=F, eye=eye, J=4)
> plot(Psi.d, axes=F, box=F, eye=eye, J=4)
```

The argument **eye** sets the viewpoint for the perspective plot (see the documentation for the **persp** function). The argument J=4 sets the sampling interval for plotting the wavelet function to 2^{-4}.

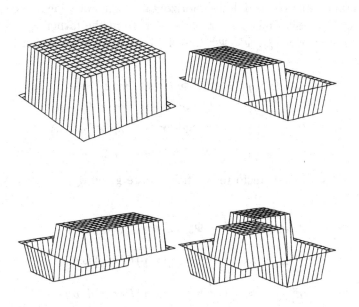

FIGURE 3.2. Plot of a family of 2-D **haar** wavelet functions Φ, Ψ^v, Ψ^h, and Ψ^d.

3.1.2   The 2-D Wavelet Approximation

We have seen how wavelets can be used to represent one dimensional (1-D) functions (2.1). The 2-D wavelet approximation is a straightforward generalization of the the 1-D approximation, although the notation gets a lot messier! You can write a 2-D function $F(x, y)$ as a sum of 2-D wavelets at different scales and locations:

$$F(x, y) \approx \sum_{m,n} s_{J,m,n} \Phi_{J,m,n}(x, y) + \sum_{j=1}^{J} \sum_{m,n} d_{j,m,n}^{v} \Psi_{j,m,n}^{v}(x, y)$$

$$+ \sum_{j=1}^{J} \sum_{m,n} d_{j,m,n}^{h} \Psi_{j,m,n}^{h}(x, y)$$

$$+ \sum_{j=1}^{J} \sum_{m,n} d_{j,m,n}^{d} \Psi_{j,m,n}^{d}(x, y). \tag{3.1}$$

Analogous to the 1-D approximation, $F(x, y)$ is decomposed into a sum of coarse resolution (level J) smooth coefficients and a sum of fine to coarse resolution (levels 1 to J) detail coefficients. However, in the 2-D approximation, there are 3 types of detail coefficients and basis functions: vertical detail, horizontal detail, and diagonal detail.

The 2-D basis functions are generated from the father wavelet Φ and mother wavelets Ψ^v, Ψ^h, and Ψ^d by scaling and translation as follows:

$$\begin{aligned}
\Phi_{J,m,n}(x, y) &= 2^{-J} \Phi(2^{-J}x - m, 2^{-J}y - n) \\
\Psi_{j,m,n}^{v}(x, y) &= 2^{-j} \Psi^v(2^{-j}x - m, 2^{-j}y - n) \\
\Psi_{j,m,n}^{h}(x, y) &= 2^{-j} \Psi^h(2^{-j}x - m, 2^{-j}y - n) \\
\Psi_{j,m,n}^{d}(x, y) &= 2^{-j} \Psi^d(2^{-j}x - m, 2^{-j}y - n).
\end{aligned}$$

The 2-D wavelet transform coefficients are given approximately by the integrals

$$\begin{aligned}
s_{J,m,n} &\approx \int\int \Phi_{J,m,n}(x, y) F(x, y)\, dx\, dy \\
d_{j,m,n}^{v} &\approx \int\int \Psi_{j,m,n}^{v}(x, y) F(x, y)\, dx\, dy \\
d_{j,m,n}^{h} &\approx \int\int \Psi_{j,m,n}^{h}(x, y) F(x, y)\, dx\, dy \\
d_{j,m,n}^{d} &\approx \int\int \Psi_{j,m,n}^{d}(x, y) F(x, y)\, dx\, dy.
\end{aligned}$$

3.2   2-D Discrete Wavelet Transform

The 2-D discrete wavelet transform (2-D DWT) computes the coefficients of the 2-D wavelet series approximation (3.1) for a $m \times n$ discrete image $\mathbf{F}_{m,n}$. The 2-D DWT maps the image $\mathbf{F}_{m,n}$ to an $m \times n$ matrix of wavelet coefficients $\mathbf{w}_{m,n}$. As in the 1-D case, a fast algorithm can be used to compute the transform; see section 12.6, Algorithms for Computing the 2-D DWT.

3.2.1   Computing the 2-D DWT

To illustrate the 2-D discrete wavelet transform, create the **xbox** test image using the **make.image** function:

```
> xbox <- make.image("xbox",nrow=128)
```

Use the **image** function to plot the xbox image

```
> image(xbox)
```

The **xbox** image is shown in figure 3.3.

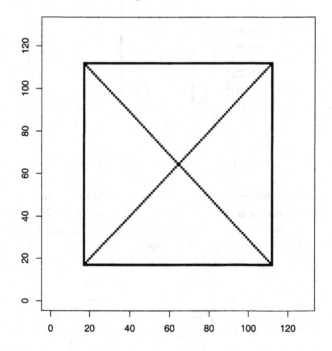

FIGURE 3.3. The **xbox** test image.

Compute and plot the 2-D DWT of the **xbox** image using the S+WAVELETS function **dwt.2d** to obtain figure 3.4:

```
> xbox.dwt <- dwt.2d(xbox, n.levels=3, wavelet="haar")
> plot(xbox.dwt)
```

In this example, the DWT is computed with **n.levels=3** resolution levels using the **haar** wavelet. The 2-D DWT is segmented

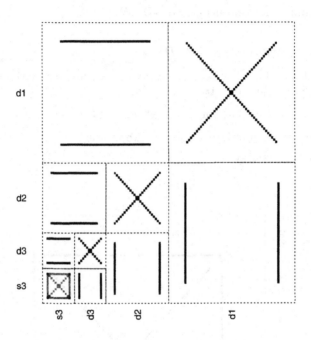

FIGURE 3.4. The DWT of the **xbox** image.

into various multiresolution coefficient matrices, corresponding to the different terms in the wavelet approximation (3.1).

Figure 3.5 shows the correspondence between the wavelet coefficients and the plot of the 2-D DWT. The relationship between the labels of figure 3.5 and the notation of the previous section is as follows:

s3–s3	$s_{3,m,n}$	(smooth)
d1–s1, d2–s2, d3–s3	$d^v_{j,m,n}$ for $j = 1, 2, 3$	(vertical detail)
s1–d1, s2–d2, s3–d3	$d^h_{j,m,n}$ for $j = 1, 2, 3$	(horizontal detail)
d1–d1, d2–d2, d3–d3	$d^d_{j,m,n}$ for $j = 1, 2, 3$	(diagonal detail)

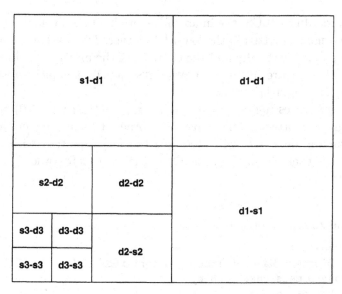

FIGURE 3.5. The 2-D DWT wavelet coefficient matrices.

The DWT of the xbox image illustrates the following:

- The s3-s3 coefficient matrix ($s_{3,m,n}$), located in the lower left-hand corner, is a fuzzy "postage stamp" replica of the full size image.

- The d3-s3, d2-s2, and d1-s1 coefficient matrices ($d^v_{3,m,n}$, $d^v_{2,m,n}$, and $d^v_{3,m,n}$), located along the x axis, correspond to the vertical edges.

- The s3-d3, s2-d2, and s1-d1 coefficient matrices ($d^h_{3,m,n}$, $d^h_{2,m,n}$, and $d^h_{3,m,n}$), located along the y axis, correspond to the horizontal edges.

- The d3-d3, d2-d2, and d1-d1 coefficient matrices ($d^d_{3,m,n}$, $d^d_{2,m,n}$, and $d^d_{3,m,n}$), located along the diagonal, correspond to the diagonal edges.

Warning: In S-PLUS, the index of the origin of an image is $(1,1)$, and the origin is plotted in the lower left corner. The rows of the matrix are plotted vertically and the columns of the matrix are plotted horizontally. As a result, in S+WAVELETS the DWT is plotted with s3-s3 in the lower left corner.

In image processing, the index of the origin is typically $(0,0)$, and the origin is plotted in the upper left corner. In many papers and books on wavelets, you will often see the DWT plotted with the coarse scale coefficients (e.g., s3-s3) in the upper left corner.

3.2.2 2-D DWT Objects

Print the `xbox.dwt` object:

```
> xbox.dwt
2D Discrete Wavelet Transform for: xbox
Image Dimensions:  128 by 128
Number of Levels:  3
Horizontal Wavelet:  haar
Vertical Wavelet:  haar
Horizontal Boundary Rule:  periodic
Vertical Boundary Rule:  periodic
Crystal names: s1-d1 d1-s1 d1-d1 s2-d2 d2-s2 d2-d2
s3-d3 d3-s3 d3-d3 s3-s3
```

The decomposition has 3 resolution levels. The **haar** wavelet and the **periodic** boundary correction rule are used for both the horizontal and vertical directions.

You can change the default settings just as in the 1-D case, except that it is possible to specify different wavelets and boundary correction rules for the horizontal and vertical directions. To find out how to do this, see the `dwt.2d` help file.

A 2-D DWT is stored as an S-PLUS matrix as depicted in figure 3.5. The S-PLUS convention for plotting matrices is to plot the rows vertically: see the warning at the end of the previous section. The matrix is stored in "column major order," which means that all elements in column i precede all elements of column $i + 1$. In terms of figure 3.5, this means the bottom row of m coefficients in the plot are the first m coefficients in storage.

In analogy with the 1-D case, the wavelet coefficient matrices are called *2-D crystals*. You can subset crystals just as in the 1-D case.

For example, you can access and plot the **s3-s3** crystal as follows to produce figure 3.6.

```
> s3.crystal <- xbox.dwt[["s3-s3"]]
> plot(s3.crystal)
```

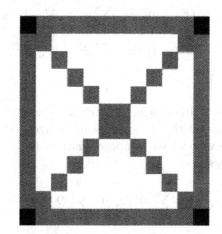

s3

s3

FIGURE 3.6. The **s3-s3** crystal for the DWT of the **xbox** image.

3.2.3   The Inverse 2-D DWT

You can invert the 2-D DWT to recover the original image. The inverse 2-D DWT is implemented with the **idwt.2d** function:

```
> xbox.recon <- idwt.2d(xbox.dwt)
```

Compute the L^2 relative error of the reconstruction:

```
> vecnorm(xbox-xbox.recon)/vecnorm(xbox)
[1] 6.055193e-16
```

See (2.11) for the definition of L^2 relative error.

Note: As in the 1-D case, to compute the inverse transform you can use the generic function **reconstruct**.

3.3 Multiresolution Analysis of Images

The 2-D functions

$$S_J(x,y) = \sum_{m,n} s_{J,m,n} \Phi_{m,n}(x,y)$$

$$D_j^v(x,y) = \sum_{m,n} d_{j,m,n}^v \Psi_{m,n}^v(x,y)$$

$$D_j^h(x,y) = \sum_{m,n} d_{j,m,n}^h \Psi_{m,n}^h(x,y)$$

$$D_j^d(x,y) = \sum_{m,n} d_{j,m,n}^d \Psi_{m,n}^d(x,y)$$

$$(3.2)$$

are called the *smooth* image, the *vertical detail* images, the *horizontal detail* images, and the *diagonal detail* images, respectively. The orthogonal 2-D wavelet series approximation (3.1) expressed in terms of these images is

$$F(x,y) \approx S_J(x,y) + \sum_{j=1}^{J} D_j^v(x,y) + \sum_{j=1}^{J} D_j^h(x,y) + \sum_{j=1}^{J} D_j^d(x,y). \quad (3.3)$$

This is a decomposition of $F(x,y)$ into $3J + 1$ orthogonal image components. As in the 1-D case, these components correspond to different resolution levels.

3.3.1 Computing a 2-D Multiresolution Decomposition

In this section, you will compute a 2-D multiresolution decomposition (2-D MRD) for the **xbox** image. Unlike the 1-D case, no function is explicitly supplied to compute the 2-D MRD. This is because the 2-D MRD is generally too large to work with in S-PLUS: see the note at the conclusion of this section. However, computing the 2-D MRD is straightforward and can easily be done using the **reconstruct** function.

Recompute the DWT for the **xbox** image using just two resolution levels and a smaller image size (64×64 instead of 128×128):

```
> xbox <- make.image("xbox", nrow=64)
> xbox.dwt <- dwt.2d(xbox, n.level=2)
```

Note: Here we use the default **s8** wavelet, rather than the **haar** wavelet, because **s8** provides a better illustration of the MRD.

Compute and plot the level 2 multiresolution image components S_2, D_2^v, D_2^h, and D_2^d to produce figure 3.7:

```
> S2.S2 <- reconstruct(xbox.dwt[["s2-s2"]])
> D2.S2 <- reconstruct(xbox.dwt[["d2-s2"]])
> S2.D2 <- reconstruct(xbox.dwt[["s2-d2"]])
> D2.D2 <- reconstruct(xbox.dwt[["d2-d2"]])
> image(sqrt(abs(S2.S2)),axes=F,xlab="S2-S2")
> image(sqrt(abs(D2.S2)),axes=F,xlab="D2-S2")
> image(sqrt(abs(S2.D2)),axes=F,xlab="S2-D2")
> image(sqrt(abs(D2.D2)),axes=F,xlab="D2-D2")
```

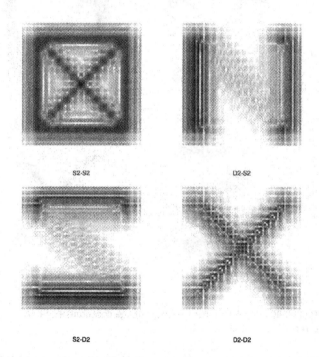

FIGURE 3.7. The level 2 multiresolution image components S_2, D_2^v, D_2^h, and D_2^d for the xbox image.

Here you used the **reconstruct** function applied to a crystal object. This gives you a reconstructed image based only that crystal.

Add up the level 2 resolution components to obtain a level 1 resolution approximation S_1 to the xbox image:

```
> S1.S1 <- S2.S2+S2.D2+D2.S2+D2.D2
```

Now compute and plot the level 1 multiresolution image components S_1, D_1^v, D_1^h, and D_1^d to produce figure 3.8:

```
> S1.D1 <- reconstruct(xbox.dwt[["s1-d1"]])
> D1.S1 <- reconstruct(xbox.dwt[["d1-s1"]])
> D1.D1 <- reconstruct(xbox.dwt[["d1-d1"]])
> image(sqrt(abs(S1.S1)),axes=F,xlab="S1-S1")
> image(sqrt(abs(D1.S1)),axes=F,xlab="D1-S1")
> image(sqrt(abs(S1.D1)),axes=F,xlab="S1-D1")
> image(sqrt(abs(D1.D1)),axes=F,xlab="D1-D1")
```

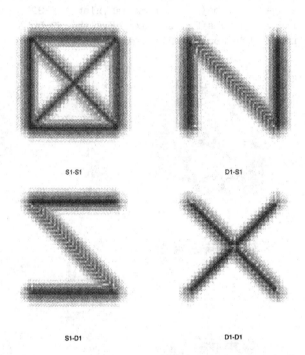

FIGURE 3.8. The level 1 multiresolution image components S_1, D_1^v, D_1^h, and D_1^d and the reconstructed xbox image.

The level 1 approximation S_1 is finer than the level 2 approximation S_2. Similarly, the level 1 detail (D_1^v, D_1^h, and D_1^d) is sharper than the level 2 detail (D_2^v, D_2^h, and D_2^d).

Note: A floating point 256×256 image in S-PLUS takes up 1 megabyte of storage. Hence, a 2-D MRD with J resolution levels of a 256×256 image takes $3J + 1$ megabytes of storage.

4

Exploratory Wavelet Analysis

One important feature of S+WAVELETS is the ability to call on the extensive functionality of S-PLUS to visualize and explore wavelet objects for "exploratory wavelet analysis." In this section, you will learn about exploratory analysis for

- 1-D DWT and other objects using the functions eda.plot and summary (4.1)

- 2-D DWT objects (4.2).

4.1 Exploratory 1-D Wavelet Analysis

There are so many things you can do in S-PLUS, it can be hard to know where to begin! For this reason, S+WAVELETS provides the function eda.plot, which selects the more interesting graphical summaries of an object. The name eda.plot comes from the nature of the plot; an eda.plot gives a graphical "exploratory data analysis." To get summaries in tabular form, you can use the function summary.

These "pre-packaged" summaries start you on a more in-depth exploration using the many S-PLUS graphical and statistical functions. For many applications, you will want to go beyond the quick pre-packaged summaries produced by the `eda.plot` and `summary` functions.

In this section, the `glint` signal is used to illustrate the statistical and visual summaries for wavelet objects. The `glint` signal represents *radar glint noise* measurements whose values are the angles of an object in degrees, corrupted by glint noise. Plot the signal as follows to obtain figure 4.1:

```
> plot(glint,type="l")
```

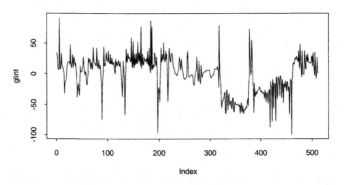

FIGURE 4.1. Radar glint noise.

The true signal is the electromagnetic "center" of a complex object. This signal is a low-frequency oscillation about 0° with possible level shift transitions. The observed signal contains a fairly large number of "glint" spikes, causing the apparent signal to be different from the true signal by as much as 150°.

4.1.1 Summaries of DWT Objects

Compute and plot the discrete wavelet transform of the glint signal:

```
> glint.dwt <- dwt(glint)
> plot(glint.dwt)
```

The DWT is plotted in figure 4.2. The large detail coefficients correspond to the glint spikes.

Now apply `eda.plot` to the `glint.dwt` object to produce figure 4.3:

```
> glint.dwt <- dwt(glint)
> eda.plot(glint.dwt)
```

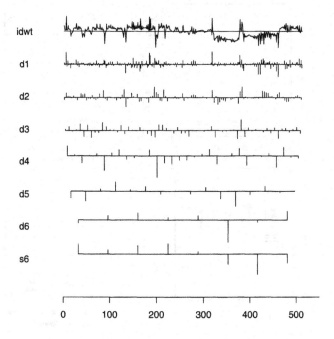

FIGURE 4.2. DWT of the glint signal using default s8 wavelet.

For DWT objects such as `glint.dwt`, the EDA plot consists of four subplots: a "time-scale plot" (top left), a "dot chart" (upper right), a "box plot" (bottom left), and an "energy plot" (bottom right). These are described below:

Time-Scale Plane Plot: The time-scale plot was discussed in section 2.4, and is produced by the `time.scale.plot` function. For the glint signal, there is significant activity in both fine and coarse scales.

Dot chart: The dot chart for a DWT object plots the percentage of energy by crystal:

$$E_j^d = \frac{1}{E} \sum_{k=1}^{n/2^j} d_{j,k}^2, \quad j = 1, \ldots, J \quad (4.1)$$

$$E_J^s = \frac{1}{E} \sum_{k=1}^{n/2^j} s_{J,k}^2 \quad (4.2)$$

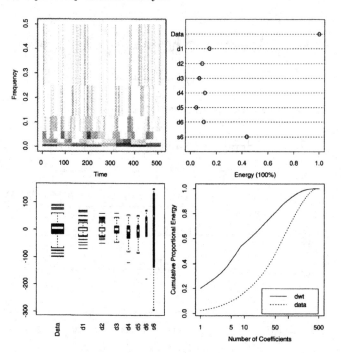

FIGURE 4.3. An EDA plot for the `glint.dwt` object. Top left: time-scale plot. Top right: percentage of energy by crystal. Bottom left: box plots of the DWT coefficients by crystal. Bottom right: distribution of energy of the DWT coefficients (solid line) and for the original data (dashed line).

where $E = \sum_{k=1}^{n} f_k^2$ is the total energy of the signal. For the orthogonal wavelets, the DWT is *energy-preserving*, so that

$$E = E_J^s + \sum_{j=1}^{J} E_j^d.$$

Dot charts are created using the `dotchart` function.

Box plot: The box plot gives a summary of the distribution of the DWT coefficients using side-by-side box plots of the data and the coefficients $\mathbf{s}_6, \mathbf{d}_6, \ldots, \mathbf{d}_1$. The boxes indicate the interquartile range and the middle white strip corresponds to the median. The width of the boxes is proportional to the square root of the number of observations in the crystal. The dynamic range tends to decrease going from coarse to fine scales. A box plot is obtained using the boxplot function.

Energy Plot: The energy plot shows the "concentration of energy." The energy concentration function for a vector $\mathbf{x} = (x_1, x_2, \ldots, x_n)'$ is defined by

$$E_x(K) = \frac{\sum_{i=1}^{K} x_{(i)}^2}{\sum_{i=1}^{n} x_i^2} \qquad (4.3)$$

where $x_{(i)}$ is the ith largest absolute value in $\mathbf{x}$. The solid line plots the energy concentration function for the wavelet coefficients $E_w(K)$. The dashed line plots $E_f(K)$, the energy concentration function for the untransformed data. By default, K is plotted on a log scale. The energy.plot function is used to create an energy plot.

From the EDA plot, we can conclude that the DWT compacts a much greater proportion of energy into fewer coefficients than the original data. Most of this energy lies in the coarse scale coefficients, particularly s_6. The energy compaction property results from the fact that wavelets are good building block functions which can represent lots of different types of signals. Turn to chapter 5 to see how this property makes wavelets valuable for applications such as nonparametric regression, signal extraction, and smoothing.

The S-PLUS generic function summary provides a tabular version of the eda.plot. Apply summary to the glint.dwt object as follows:

```
> print(summary(glint.dwt),digits=2)
        Min      1Q Median     3Q
s6 -296.35 -132.45  49.40 128.09
d6 -180.31  -19.33  27.77  40.95
d5  -85.78  -23.13  -2.96  14.06
d4 -119.03  -27.23  -3.45  13.91
d3  -46.33  -12.32   1.35  13.79
d2  -49.59   -5.13   0.34   7.42
d1  -69.93   -5.09   0.29   7.15
        Max  Mean     SD    MAD Energy.%
s6 150.67 -6.59 164.44 146.74     0.44
```

```
d6   71.87 -0.24  80.05  55.13      0.10
d5   51.93 -8.20  34.97  28.81      0.04
d4   54.92 -8.06  38.52  29.36      0.11
d3   59.79  0.16  21.71  18.57      0.07
d2   59.29  2.46  17.32   9.58      0.09
d1   70.12  0.85  15.77   8.38      0.15
```

```
Energy Distribution:
              1st     1%     2%     3%
Energy.%     0.20   0.47   0.58   0.63
|coeffs|   296.35 127.89  71.87  61.89
#.coeffs     1.00   6.00  11.00  16.00
              4%     5%    10%    15%
Energy.%     0.67   0.70   0.82   0.88
|coeffs|    58.03  52.13  36.64  28.93
#.coeffs    21.00  26.00  52.00  77.00
             20%    25%
Energy.%     0.92   0.94
|coeffs|    22.27  18.97
#.coeffs   103.00 128.00
```

For each crystal, summary computes the smallest and largest coefficients (Min and Max), the lower and upper quartiles (1Q and 3Q), the median, the mean, the standard deviation (SD), a robust estimate of statistical scale of the data (MAD, which is 1.48 times the median absolute deviation from the median), and the proportion of energy in the original signal accounted for by that crystal.

The summary function also computes the distribution of the energy of the coefficients. For the glint signal, the largest coefficient accounts for 20% of the total energy and the largest six (1%) coefficients account for 47% of the total energy.

4.1.2 Summaries of Decomposition Objects

Use eda.plot on a decomposition object to create figure 4.4:

```
> glint.decomp <- mrd(glint)
> eda.plot(glint.decomp)
```

For decomposition objects, an EDA plot produces a box plot and a dot chart. The dot chart for the decomposition object is identical to that for the coefficients; compare with figure 4.3. This is because the energy of the signal components is the same as the energy of the coefficient crystals.

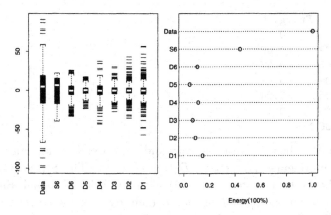

FIGURE 4.4. An EDA plot for the `glint.decomp` object. Left: box plots by signal component. Right: proportion of energy represented by each signal component.

The summary function provides a statistical summary of the object `glint.decomp`:

```
> print(summary(glint.decomp),digits=2)
```

```
Correlation Matrix:
     Data    D1   D2    D3    D4    D5    D6    S6
     1.00  0.38  0.3  0.26  0.33  0.21  0.32  0.66
D1   0.38  1.00  0.0  0.00  0.00  0.00  0.00  0.00
D2   0.30  0.00  1.0  0.00  0.00  0.00  0.00  0.00
D3   0.26  0.00  0.0  1.00  0.00  0.00  0.00  0.00
D4   0.33  0.00  0.0  0.00  1.00  0.00  0.00  0.00
D5   0.21  0.00  0.0  0.00  0.00  1.00  0.00  0.00
D6   0.32  0.00  0.0  0.00  0.00  0.00  1.00  0.00
S6   0.66  0.00  0.0  0.00  0.00  0.00  0.00  1.00
```

```
Variances:
    Data      D1     D2     D3    D4 D5     D6      S6
  848.96  124.49  76.06  58.12  94.07 38  87.78  370.44
```

```
Statistics for Components:
        Min      1Q Median     3Q    Max   Mean     SD    MAD
D1  -56.92   -4.38  -0.03   3.49  56.72   0.00  11.16   5.88
D2  -35.43   -3.06  -0.09   3.23  43.26   0.00   8.72   4.65
D3  -26.95   -4.02  -0.07   3.74  30.49   0.00   7.62   5.71
D4  -42.88   -5.21  -0.15   5.87  37.53   0.00   9.70   8.23
D5  -23.16   -3.45   0.37   3.14  17.70   0.00   6.16   4.98
D6  -33.07   -5.23  -0.35   5.67  25.97   0.00   9.37   7.99
S6  -39.50  -17.13   6.53  15.41  22.39  -0.82  19.25  17.34
```

```
        Energy.%
   D1     0.15
   D2     0.09
   D3     0.07
   D4     0.11
   D5     0.04
   D6     0.10
   S6     0.44
```

This gives the correlation matrix and variances for the data and signal components $S_6(t), D_6(t), D_5(t), \ldots, D_1(t)$. The $S_6(t)$ component has the highest correlation (.65) with the original signal. Since we are using an orthogonal wavelet decomposition, the correlation matrix of the components $S_6(t), D_6(t), D_5(t), \ldots, D_1(t)$ is diagonal. For non-orthogonal transforms, this may not be the case (see section 2.5).

The summary function also gives a table of summary statistics of the signal components. These are the same as for DWT objects (see the discussion on page 58).

4.1.3 Summaries of Crystals

The EDA plot and summary for individual crystals gives a more detailed look at that crystal. For example, the EDA plot for the d1 wavelet coefficient vector

```
> eda.plot(glint.dwt[["d1"]])
```

is shown in figure 4.5. This EDA plot helps in assessing the autocorrelation and distribution of the wavelet coefficients. This information is important for applications such a statistical estimation and data compression.

The four subplots of this EDA plot are:

Stack Plot: The stack plot gives the coefficients d1 plotted with the reconstructed component signal D1. This helps show the role coefficients play in representing the original signal.

ACF Plot: The autocorrelation function (ACF) of the $\mathbf{d}_1$ wavelet coefficients. The dashed lines are 95% confidence intervals for the autocorrelations to be "significantly" different from zero. Here the $\mathbf{d}_1$ coefficients do not exhibit significant autocorrelations.

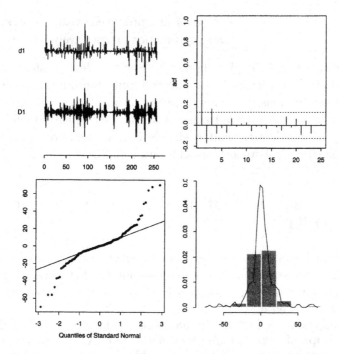

FIGURE 4.5. An EDA plot for the **d1** wavelet coefficient vector.

Q-Q Plot: Quantile-Quantile plot of the wavelet coefficients versus the quantiles of a standard normal. If the points follow the line in this plot, then you can infer that the distribution is normal. Here you can see that the distribution is quite non-normal. The curvature of the points indicates that the coefficients have heavy (long) tails. This plot is produced using the **qqnorm** function.

Histogram and Density Plot:

      The histogram and an estimate of the density function give another way to visualize the distribution of the coefficients. For the **d1** coefficients in this example, these plots confirm that the distribution has longer tails than a normal distribution. The histogram and density plots are produced with the **hist** and **density** functions respectively.

Note: The ACF plot shows that the wavelet transform does a good job of "whitening," or uncorrelating, the input data for the **glint** signal. This is a general feature of wavelets, which have been shown to whiten a broad class of signals; see [TK92, Fla92]. The ability of wavelets to whiten signals is important for applications such as statistical smoothing and data compression.

4.2   Visual and Statistical Summaries for the 2-D DWT

The visual and statistical summaries which you produced for 1-D wavelets (see section 4.1) can also be done for 2-D wavelets. In this section, you will learn what the EDA plot and summary functions do in the 2-D case.

As an example, look at the **daubechies** image, which is a digital photograph of Ingrid Daubechies at the 1993 AMS winter meetings in San Antonio, Texas. Ingrid Daubechies is a very active researcher in the field of wavelet analysis and author of the book *Ten Lectures on Wavelets* [Dau92]. She is the inventor of smooth orthonormal wavelets of compact support, which include the "d4" wavelet. The photograph was taken by David Donoho, another leading wavelets researcher, with a Canon XapShot video still frame camera. Plot the **daubechies** image and take its DWT to produce figure 4.6 as follows:

```
> daub.dwt <- dwt.2d(daubechies, n.level=3)
> par(mfrow=c(1,2))
> image(daubechies)
> plot(daub.dwt)
```

To enhance the detail coefficients, the **s3-s3** coefficients are plotted on a different scale. To enhance the coefficients at fine scales, the logarithms of the absolute values of the detail coefficients are plotted instead of the actual coefficients. To plot the absolute values of the raw coefficients, use the **abs** function and set the argument **stretch.image=F** to produce figure 4.7:

```
> plot(abs(daub.dwt),stretch.image=F)
```

Compute the EDA plot of the **daub.dwt** object to produce figure 4.8:

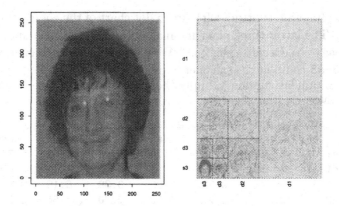

FIGURE 4.6. The **daubechies** image and its 2-D discrete wavelet transform.

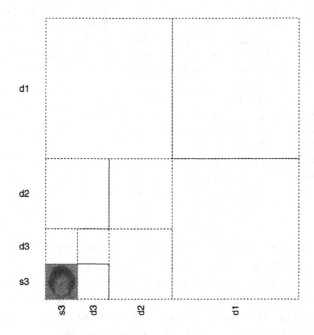

FIGURE 4.7. The raw coefficients for the 2-D discrete wavelet transform of the **daubechies** image.

```
> eda.plot(daub.dwt)
```

The EDA plot consists of four subplots: the location of the top 5% of the wavelet coefficients (top left), side-by-side box plots of the wavelet coefficients grouped by crystal (top right), the energy concentration function (bottom left), and a dot chart of the energy (bottom right). The latter three plots are analogous to their 1-D counterparts (see page 55). Because the 2-D DWT compacts so much energy into the s3-s3 component, the energy plot is done on a log-log scale.

For a tabular summary, apply the summary function to the object daub.dwt:

```
> print(summary(daub.dwt),digits=2)
            Min       1Q  Median       3Q      Max     Mad
s1-d1    -67.75    -0.76   -0.24     0.28    11.52    0.77
d1-s1    -20.37    -2.32   -0.28     1.82    29.18    3.07
d1-d1     -5.22    -0.38    0.00     0.39     6.12    0.57
s2-d2   -121.69    -3.22   -0.02     2.90   104.56    4.51
d2-s2    -97.99    -4.08   -0.37     3.17   126.81    5.35
d2-d2    -32.73    -1.87   -0.04     1.89    45.53    2.80
s3-d3   -181.65    -8.55    0.49     8.35   288.85   12.64
d3-s3   -276.71    -7.95    0.47     7.68   310.42   11.38
d3-d3   -117.95    -4.16   -0.10     5.10   147.45    6.87
s3-s3    547.96   749.51  829.69  1273.03 1494.58  180.95
            Mean       SD Energy.%
s1-d1     -0.36     2.57        0
d1-s1     -0.24     3.49        0
d1-d1      0.00     0.59        0
s2-d2     -0.36    10.64        0
d2-s2     -0.61    11.74        0
d2-d2     -0.03     4.98        0
s3-d3     -0.52    31.25        0
d3-s3      0.35    37.72        0
d3-d3      0.32    17.59        0
s3-s3    972.93   272.93        1
```

Almost all of the energy of the discrete wavelet transform is captured by the coarsest smooth coefficients—those in s3-s3. The minimum of the coefficients in s3-s3 is about 548 and the maximum is over 1494. The maximum coefficient in absolute value at all other scales is less than 311. The coefficients are smallest in magnitude at the finest scales d1-s1, s1-d1, and d1-d1, and increase in magnitude at the coarser scales. In fact, even though the d1-d1 component represents 25% of the total coefficients, not a single d1-d1 coefficient is in the top 5%. The largest detail coefficients correspond to edges in the original image. We also see some large coefficients associated with boundary effects.

Apply the EDA plot to a crystal to get a more detailed look at a particular DWT coefficient component. For example, to create fig-

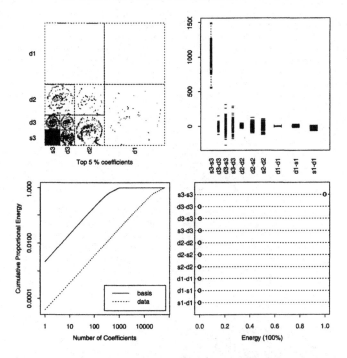

FIGURE 4.8. EDA plot for the 2-D DWT of the **daubechies** image. Top left: location of the top 5% of the coefficients. Top right: side-by-side box plots of the wavelet coefficients grouped by crystal. Bottom left: the energy concentration function for the wavelet coefficients and the original image. Bottom right: a dot chart of the energy by crystal.

ure 4.9, which is an EDA plot of the **s2-d2** crystal, extract the crystal inside the **eda.plot** function as follows:

```
> eda.plot(daub.dwt[["s2-d2"]])
```

The 2-D EDA plot is analogous to the EDA plot for a 1-D crystal (see page 61). The coefficient matrix **s2-d2** (top left) is displayed alongside the the reconstructed image component **S2-D2** (top right). The **S2-D2** component picks up horizontal edges. The 2-D autocorrelation function (bottom left), indicates the degree of spatial correlation present in the coefficient matrix. In this case, there is non-negligible correlation in the horizontal direction. The histogram and density estimate (bottom right) shows that the distribution of the coefficient matrix **s2-d2** is long-tailed and much more concentrated than a normal distribution.

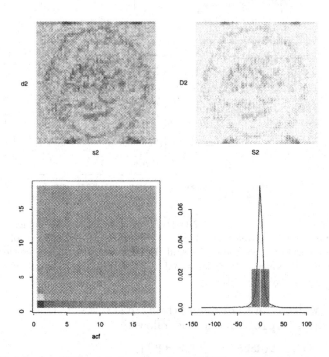

FIGURE 4.9. An EDA plot for the s2-d2 DWT coefficient matrix of the daubechies image. Top left: the coefficient matrix s2-d2. Top right: the reconstructed image component S2-D2. Bottom left: the 2-D autocorrelation function of the coefficient matrix. Bottom right: the histogram and density estimate for the coefficients.

5
More on Wavelet Analysis

An important feature of the S-PLUS language is the flexibility of the data structures. You can interactively examine, visualize, and modify wavelet objects. In this chapter, you will explore wavelet analysis in more depth, learning how to perform the following tasks:

- Select options for your wavelet analysis using arguments to the dwt, mrd, and mra functions and the wavelet.options function (section 5.1).

- Extract individual crystals from a DWT structure in different ways (section 5.2).

- Extract multiple crystals from a DWT structure to form *wavelet packets* (section 5.3).

- Use logical and math operators on DWT objects (section 5.4).

- Replace parts of a DWT structure with a constant value or a vector of values (section 5.5).

- Access and replace individual coefficients (section 5.6).

- Operate directly on the DWT object, bypassing the specialized subscripting and replacement operators (section 5.7).

5.1 Selecting Options for Wavelet Analysis

In performing a wavelet analysis of a signal, you can adjust three primary optional arguments:

n.levels: The number of multiresolution levels.

wavelet: The wavelet family.

boundary: The boundary correction rule.

For most applications, you can use the default boundary correction rule. Turn to chapter 14, Boundary Conditions for Wavelet Analysis, to learn about selecting a non-default boundary correction rule. In this section, you will learn about selecting the number of multiresolution levels and the wavelet family.

5.1.1 Selecting the Multiresolution Level

The number of multiresolution levels in a wavelet analysis can be set using the argument n.levels. By default, n.levels is set to the largest possible integer less than or equal to 6 (it is not always possible to compute 6 levels).

In some applications, you may want to set n.levels to be less than 6 levels. An example can be found on page 30, in which the multiresolution analysis was restricted to 4 levels. In that example, "zooming" to coarser scales does not reveal any useful information.

In other applications, you will want to set n.levels to be greater than 6. This is particularly true when you have a signal with more than 2048 sample values. For example, more than 6 levels is often helpful for estimating the rate of decay in the wavelet coefficients (the rate of decay can be used to characterize features in a signal, such as singularities [MH92, MZ92]). It is important to use enough multiresolution levels to obtain an accurate estimate. To illustrate this, create the plot of figure 5.1 as follows:

```
> par(mfrow=c(1,2))
> ramp <- make.signal("ramp", n=8192)
> plot(dwt(ramp, n.level=4))
> plot(dwt(ramp, n.level=8))
```

The decay in the wavelet coefficients is visually more pronounced in the DWT with 8 levels than in the DWT with 4 levels.

To obtain the maximum number of levels, you would set the argument n.levels=NULL. The maximum number of levels depends on

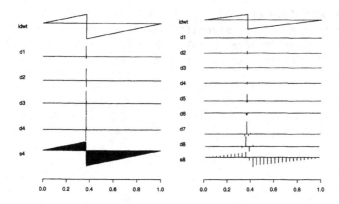

FIGURE 5.1. DWT of the ramp signal of length $n = 8192$. Left: using n.levels=4. Right: using n.levels=8.

the sample size, boundary condition, and type of wavelet (the number of coefficients at the coarsest level must be at least one half the length of the wavelet filter). You can use the function max.level to determine the maximum level.

Note: You can also set the default values for n.levels, wavelet, and **boundary** using the function wavelet.options. Instead of passing arguments to the dwt, mrd, and mra functions, you can reset the default values. See the wavelet.options help file for details.

5.1.2   Selecting a Wavelet

There are no hard and fast rules for selecting a wavelet to use for an analysis. By default, the functions dwt, mrd, and mra use the s8 wavelet. This is a good overall choice for many applications: the s8 wavelet is orthogonal, smooth, nearly symmetric, and non-zero on a relatively short interval (short support). You can change the default wavelet using the optional argument wavelet. Table 2.1 on page 17 lists other wavelets you can use.

A central reason to use a particular wavelet is to match the characteristics of the signal you are analyzing. To illustrate this concept, compare the haar and s8 multiresolution analyses for the blocks signal, which is a piecewise constant function. Produce figure 5.2 as follows:

```
> par(mfrow=c(1,2))
> blocks <- make.signal("blocks", n=1024)
> plot(mra(blocks, wavelet="s8"))
> plot(mra(blocks, wavelet="haar"))
```

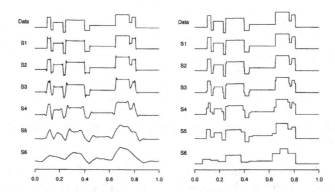

FIGURE 5.2. Multiresolution analysis of the blocks signal. Left: using the s8 wavelet. Right: the haar wavelet.

The s8 wavelet approximations, which are smooth by nature, tend to blur the discontinuities. By contrast, the haar wavelet, which itself is a piecewise constant function, preserves the discontinuities. For the blocks signal, the haar wavelet provides a much better basis for analysis.

However, for a smooth signal, the haar wavelet results in an unnecessarily non-smooth analysis. Compare a multiresolution analysis of the doppler signal for the s8 wavelet and the haar wavelet, as shown in figure 5.3:

```
> par(mfrow=c(1,2))
> doppler <- make.signal("doppler", n=1024)
> plot(mra(doppler, wavelet="s8"))
> plot(mra(doppler, wavelet="haar"))
```

Which wavelet to choose is usually not so clear cut as in these examples. There are many different factors which you may want to take into account. Two important factors are the smoothness and the spatial localization of the wavelet. In general, the wavelets with wider support (those with a big number in table 2.1) are smoother but are spatially less localized.

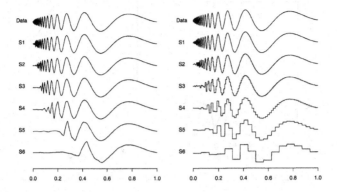

FIGURE 5.3. Multiresolution analysis of the **doppler** signal. Left: using the **s8** wavelet. Right: the **haar** wavelet.

There are also other properties of the wavelet which may be important. You will often have to trade off competing goals. Turn to section 12.4, Properties of Wavelet Functions, to learn more about the properties of wavelet functions and the tradeoffs which must be made.

You can also reverse the wavelet filters in a wavelet analysis. The reversed wavelet filters have the same orthogonality properties and smoothness as the original wavelet filters. Analysis with reversed filters is equivalent to analysis with the wavelets reversed in the time domain. On both theoretical and practical levels, using reversed wavelet filters makes very little difference. The following example shows a **d4** father wavelet (top left), the reversed **d4** father wavelet (top right) and the DWTs of the **doppler** signal using the standard **d4** filter (bottom left) and the reversed **d4** filter (bottom right).

```
> d4.filt <- wave.filter("d4")
> doppler <- make.signal("doppler")
> dopp.d4 <- dwt(doppler, wavelet="d4", n.levels=4)
> dopp.d4.rev <- dwt(doppler, wavelet="d4", n.levels=4,
+     filter.reverse=T)
> par(mfrow=c(2,2))
> plot(wavelet("d4", mother=F))
> plot(wavelet("d4.rev", filter=rev(d4.filt), mother=F))
> plot(dopp.d4)
> plot(dopp.d4.rev)
```

The result is displayed in figure 5.4.

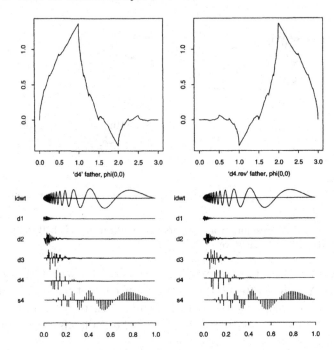

FIGURE 5.4. Top left: the **s8** father wavelet. Top right: the reversed **s8** father wavelet. Bottom left: the DWT of the `doppler` signal using **s8** filter. Bottom right: the DWT of the `doppler` signal using the reversed **s8** filter.

5.2   Extracting Crystals from a DWT Object

A DWT object is a special vector of wavelet coefficients stored in a certain order. The ordering is described in section 2.2.2. Along with the vector are *attributes* which describe the wavelet transform.

5.2.1   Extracting Crystals from 1-D DWT Objects

As described in section 2.2.2, the simplest way to extract a crystal from a one-dimensional DWT object is to use the [[operator. For example, suppose you create and print a DWT object using the `doppler` signal:

```
> doppler <- make.signal("doppler", n=1024)
> doppler.dwt <- dwt(doppler)
> doppler.dwt
Discrete Wavelet Transform of doppler
Wavelet: s8
Length of series: 1024
```

```
Number of levels: 6
Boundary correction rule: periodic
Crystals: s6 d6 d5 d4 d3 d2 d1
```

To extract the first crystal s6, you can subscript by crystal name, as in the expression `doppler.dwt[["s6"]]`:

```
> doppler.dwt[["s6"]]
      s6(1)          s6(2)          s6(3)      s6(4)        s6(5)
 0.007765673 -0.001330175 -0.07550149 0.110322 0.1329815
      s6(6)        s6(7)      s6(8)        s6(9)      s6(10)
 -0.466046 -0.6727822 3.648829 -1.715457 -4.011206
     s6(11)     s6(12)     s6(13)    s6(14)     s6(15)      s6(16)
 -1.536702 1.623148 3.211735 3.10738 2.040245 0.7879406
```

You can also subscript by the "internal" crystal name. This is the name actually stored in the data structure, and is different from the external name that gets printed out. The internal name is based on the sequence of filtering operators applied to obtain that crystal; see chapter 12. Obtain the internal names using the function `crystal.names`:

```
> crystal.names(doppler.dwt)
 LLLLLL HLLLLL HLLLL HLLL  HLL   HL    H
  "s6"   "d6"   "d5"  "d4"  "d3"  "d2"  "d1"
attr(, "is.on"):
 orthogonal complete
        T          T
```

The internal names are listed on top of the usual names. Thus, an alternative way to extract the s6 crystal is with the expression `doppler.dwt[["LLLLLL"]]`.

Note: The attribute `is.on` indicates whether the crystal names correspond to a complete and orthogonal basis for a signal. You can also use the function `is.on.basis` to determine whether a set of names corresponds to a basis. This function is particularly useful for wavelet packet and cosine packet analysis.

In addition to subscripting by name, you can subscript by number. The crystals are stored in the order indicated by the printed DWT object or by the `crystal.names` function. Hence, a third way to extract the s6 crystal is `doppler.dwt[[1]]`.

5.2.2 Extracting Crystals from 2-D DWT Objects

Extracting crystals from two-dimensional (2-D) objects is very similar to the one-dimensional case. You can extract crystals by name, number, or internal name. The 2-D internal names can be obtained using the crystal.names function.

The subscript operator [[produces a 2-D crystal matrix. The subscript operator [can be used to extract multiple crystals; it returns a dwt.2d object of the same dimensions as the original matrix. Zeros are inserted for the coefficients which were *not* extracted.

5.3 Creating Wavelet Packets

When you extract more than a single crystal from a DWT object, the resulting object is no longer a crystal. Rather, it is a *wavelet packet* object. To create a wavelet packet from a DWT object, use the [operator with a vector of crystal names. For example, extract the level 6 resolution crystals from the doppler.dwt object as follows:

```
> lev6 <- doppler.dwt[c("s6", "d6")]
```

This creates a new vector lev6 consisting of the coefficients from just two crystals, s6 and d6. Plot the reconstruction from lev6 to produce figure 5.5:

```
> plot(reconstruct(lev6),type="l")
```

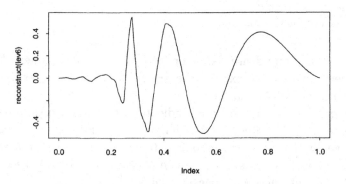

FIGURE 5.5. The reconstruction from the lev6 object.

The reconstructed signal is exactly the same as the s5 multiresolution approximation given in figure 2.10.

The object lev6 is a wavelet packet object which inherits from the class wpt. It does not inherit from the class dwt, since it does not form a basis and you cannot obtain the original signal. The wavelet packet class includes a more general class of transforms, discussed in chapter 7. For the most part, you can treat lev6 just as you would

a DWT object, keeping in mind that `lev6` is not a complete basis
for the original signal.

Note: The operator `[` behaves differently in the 1-D and 2-D
cases. In the 1-D case, it returns a smaller vector including only
the subscripted coefficients. In the 2-D case, it returns the entire
coefficient matrix structure with the non-subscripted crystals set to
zero.

5.4 Math and Logical Operators Applied to DWT Objects

In general, you can apply math and logical operators to DWT objects just as you would to ordinary S-PLUS vectors and matrices.
For example, you can take the DWT of the `glint` signal and compute the median absolute deviation from the median (MAD) of the
coefficients:

```
> glint.dwt <- dwt(glint)
> glint.mad <- mad(glint.dwt)
> glint.mad
[1] 11.22073
```

The MAD is a robust estimate of scale. Create and plot a logical
vector indicating which coefficients are bigger than 2 times the MAD:

```
> plot(abs(glint.dwt) > 2*glint.mad, data=glint,
+       zerocenter=T)
```

The result is shown in figure 5.6, in which the coefficients are plotted
either as a zero (less than 2 times the MAD) or as a one (greater than
2 times the MAD). The largest coefficients are concentrated at the
coarse scales and near the glint spikes. The argument `data=glint` ensures that the top row is the data rather than the IDWT of the zero-
one DWT structure. The argument `zerocenter` adjusts the scaling
on the y-axis, forcing each level to be centered at zero.

5.5 Replacement in DWT Objects

Just as you can access individual crystals, it is possible to replace
the coefficients in individual crystals with different numeric values.
For example, to set all the `d1` coefficients to zero, use replacement
with the `[[` operator as follows:

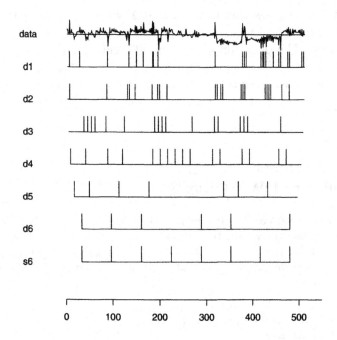

FIGURE 5.6. The location of the wavelet coefficients which are bigger than 2 times the MAD.

```
> doppler.dwt[["d6"]] <- 0
```

To set the coefficients of more than one crystal to zero, use replacement with the [operator:

```
> doppler.dwt[c("d1","d2","d3")] <- 0
```

Now plot doppler.dwt to produce figure 5.7:

```
> plot(doppler.dwt)
```

The IDWT of doppler.dwt, plotted in the top row of figure 5.7, produces the multiresolution component $S_3(t)$.

You can also replace crystals using vectors of values. This is useful, for example, for simulation of multiscale processes; see section 5.5.2 below. The replacement operators for 2-D objects are identical to the 1-D operators.

5.5.1 Replacement Using Logical Vectors

When performing replacement with [, you can use a logical vector of the same length as the number of wavelet coefficients as a subscript.

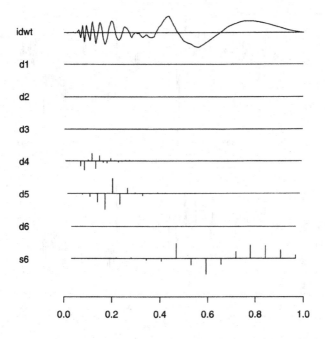

FIGURE 5.7. The `doppler.dwt` object after the **d1**, **d2**, and **d3** have been set to zero.

This is very useful to set certain coefficients to zero. For example, to reconstruct from the glint wavelet coefficients larger than 2 times the MAD, you can do the following;

```
> glint.dwt <- dwt(glint)
> glint.mad <- mad(glint.dwt)
> glint.dwt[abs(glint.dwt) < 2*glint.mad] <- 0
> glint.recon <- reconstruct(glint.dwt)
```

Compare the reconstructed signal with the original signal as follows to produce figure 5.8:

```
> par(mfrow=c(2,1))
> plot(glint, type="l")
> plot(glint.recon, type="l")
```

Note: You can also subscript a wavelet object with a logical vector having the same length as the number of coefficients. This produces a *molecule* object containing a subset of coefficients. It is called a molecule because the coefficients may be in any order, and are not necessarily organized on a lattice (as with crystals). You can

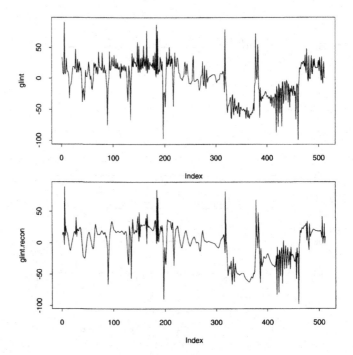

FIGURE 5.8. The original **glint** signal and the reconstructed **glint** signal based on coefficients larger than 2 times the MAD.

find out more about molecule objects in section 10.3.

5.5.2   Simulation of Multiscale Processes

You can use the assignment operator to generate random multiscale processes. For example, try simulating a series which has normal random DWT coefficients at the finest detail scale. To do this, first create a dummy DWT structure with all zero coefficients, assign the d1 crystal to random noise, and reconstruct:

```
> zero.dwt <- dwt(rep(0,512))
> zero.dwt[["d1"]] <- rnorm(256)
> series1 <- reconstruct(zero.dwt)
```

Now generate a series which has normal random DWT coefficients at the resolution levels 2 and 3:

```
> zero.dwt <- zero.dwt*0
> zero.dwt[c("d2","d3")] <- rnorm(128+64)
> series2 <- reconstruct(zero.dwt)
```

Multiplying the `zero.dwt` object by 0 is a fast way to convert an existing DWT object to one with all zero coefficients. The two series and their autocorrelation functions are compared in figure 5.9, which is produced as follows:

```
> par(mfrow=c(2,2))
> plot(series1,type="l")
> acf(series1)
> plot(series2,type="l")
> acf(series2)
```

Both `series1` and `series2` are normally distributed time series with degenerate covariance matrices.

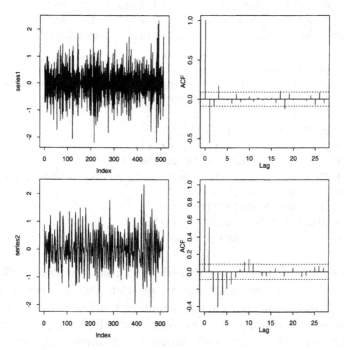

FIGURE 5.9. Two simulated multiscale random process and their autocorrelation functions (acf's).

5.6   Subscripting by Coefficient

Thus far, you have learned how to subset and replace entire crystals. To access or assign individual coefficients of a DWT object, you need to first access the crystal and then subset or assign the coefficients within that crystal. This section illustrates this process using the *cascade algorithm*, which provides one way to plot a wavelet function. The cascade algorithm is discussed in [Dau92] (p. 202) and [Str89] ("construction 1").

The cascade algorithm is essentially the same as the backward pyramid algorithm for computing the IDWT, except that the wavelet filters are reversed. The algorithm starts from a vector consisting of all zeros, except for a single value equal to one, and the wavelet function can be approximated by recursive refinement.

You can implement the cascade algorithm by creating a DWT object using the appropriate wavelet from a zero vector and then setting a coarse scale coefficient to one. For example, produce two different plots of the d4 wavelet in figure 5.10 as follows:

```
> par(mfrow=c(1,2))
> # First run of the cascade algorithm: interior wavelet
> zero.dwt <- dwt(rep(0,256), wavelet="d4",
+       filter.reverse=T)
> zero.dwt[["s6"]][2] <- 1
> plot(idwt(zero.dwt), type="l",
+       ylab="interior d4 wavelet")
> # Second run of the cascade algorithm: boundary wavelet
> zero.dwt[["s6"]] <- 0
> zero.dwt[["s6"]][1] <- 1
> plot(idwt(zero.dwt), type="l",
+       ylab="boundary d4 wavelet")
```

The first run of the cascade algorithm, in which a middle coefficient is set to one, produces an "interior" wavelet. The second run of the cascade algorithm, in which a boundary coefficient is set to one, produces a "boundary" wavelet. Since the default boundary rule in S+WAVELETS is periodic, the boundary d4 wavelet is a periodized version of the interior wavelet.

Note: The plot function in S+WAVELETS uses a different technique to plot and evaluate wavelet functions. This technique is numerically more stable than the cascade algorithm: see section 12.5.

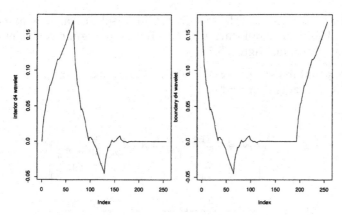

FIGURE 5.10. Illustration of the cascade algorithm. Left: a plot of a father wavelet **d4** function generated by the cascade algorithm. Right: a plot of a boundary **d4** wavelet.

5.7 Bypassing the DWT Subscript, Assignment, and Math Operators

The usual subscript, assignment, and math operators treat DWT objects as a set of crystals and decompose objects as a set of signal components. If you want to treat a DWT object or decompose object as a simple vector or matrix, you need to first coerce the object removing the class structure and then apply the default S-PLUS math and subscripting operators. In this way, you can bypass the special DWT subscript and math operators and directly access the coefficients. This requires, however, that you know the index of the coefficients in which you are interested. The ordering of the 1-D DWT and 2-D DWT coefficients is discussed in sections 2.2.2 and 3.2.2.

As an example, suppose that you want to compute the threshold corresponding to the p largest coefficients. To do this, first coerce your DWT object as a simple $1 \times n$ matrix using the **as.matrix** function:

```
> glint.dwt <- dwt(glint)
> glint.mat <- as.matrix(glint.dwt)
> glint.100 <- rev(sort(abs(glint.mat)))[100]
> glint.100
[1] 23.04094
```

The 100*th* largest wavelet coefficient in absolute value is given by
`glint.100`. Plot the location of the 100 largest glint coefficients as
follows to produce figure 5.11:

```
> plot(abs(glint.dwt) > glint.100, data=glint,
+       zerocenter=T)
```

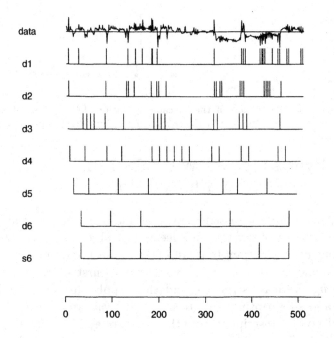

FIGURE 5.11. The location of the 100 largest wavelet coefficients in absolute
value.

To subset **glint.dwt** as a single long vector, **glint.dwt** is coerced
to a matrix in this example.

Note: There are other ways to remove the class information from
an object. You can also use the **as.vector** or **unclass** functions.
The **as.matrix** function has the advantage of producing descriptive
labels naming the elements of the vector. This can help to identify
the parts of the vector you want.

6

Nonparametric Estimation with Wavelets

One of the great success stories of wavelets is in the field of nonparametric statistical estimation. The use of wavelets in statistical applications was pioneered by professors David Donoho and Iain Johnstone at Stanford University. Their methodology rests upon the principle of *wavelet shrinkage*, which refers to removing noise (denoising) by shrinking wavelet coefficients towards zero. In this chapter, you will learn about the following topics:

- Using wavelets and wavelet shrinkage for nonparametric regression and smoothing with the `waveshrink` function (section 6.1).

- The principles behind wavelet shrinkage for nonparametric statistical estimation (section 6.2).

- Assessing the wavelet shrinkage estimates using `summary` and the various diagnostic plots produced by `eda.plot` (section 6.3).

- Tuning the wavelet shrinkage estimates by adjusting different arguments to the `waveshrink` function (section 6.4).

- Comparing the wavelet shrinkage estimates with other nonparametric regression estimators and smoothers (section 6.5).

- Applying the bootstrap method to WaveShrink in order to obtain variance estimates (section 6.7).

6.1   Nonparametric Regression and Smoothing

In nonparametric regression problems, you want to estimate an unknown signal $f(t)$ from some noisy data y_i. Suppose your data are
given by

$$y_i = f_i + \epsilon_i$$

where f_i is a discrete signal and ϵ_i are independent and identically
distributed normal errors: $\epsilon_i \sim N(0, \sigma^2)$. The usual parametric regression requires that you form a particular model for f_i (e.g., f_i are
samples of a piecewise polynomial). In nonparametric regression, you
make minimal assumptions about the exact nature of f_i.

Donoho and Johnstone [Don93a, DJ94, DJKP95, Don95, DJ95]
have developed an impressive theory and methodology for nonparametric regression and smoothing based on the principle of *wavelet
shrinkage*. Wavelet shrinkage refers to estimates obtained by:

1. Applying the discrete wavelet transform.

2. Shrinking the wavelet coefficients towards zero.

3. Applying the inverse discrete wavelet transform.

Following Donoho and Johnstone, the wavelet shrinkage methodology is referred to here as WaveShrink.

The Donoho and Johnstone WaveShrink estimators are implemented by the function `waveshrink`. We try out the `waveshrink`
function on the signal nmr1, a nuclear magnetic resonance (NMR)
signal of length 1024. Plot the signal as follows to produce figure 6.1.

```
> plot(nmr1, type="l", ylab="NMR Spectrum")
```

The NMR signal has several pronounced features, including a large
spike about halfway through the series. However, the observed signal
exhibits a fair amount of apparent background noise. You would
like to remove the background noise from the NMR signal without
blurring or removing the features such as the large spike.

Applying `waveshrink` to nmr1 and plotting the result yields figure 6.2:

```
> nmr1.sm <- waveshrink(nmr1)
> plot(nmr1.sm, type="l", ylab="Waveshrink Estimate")
```

You can see that `waveshrink` does a good job of removing the noise
while preserving the features.

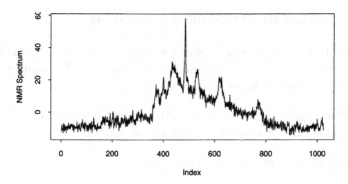

FIGURE 6.1. Plot of nmr1 spectrum signal.

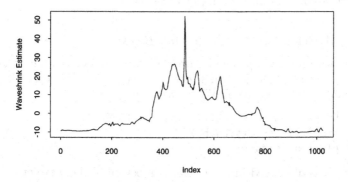

FIGURE 6.2. WaveShrink estimate of the nmr1 signal.

Note: The methodology for wavelet shrinkage nonparametric estimation shows great promise. Many new methods for wavelet shrinkage estimation are being developed. Nason [Nas95] has proposed a version of WaveShrink which uses cross-validation. Ogden [OP94] has developed a new "data-dependent" thresholding scheme for WaveShrink. Several researchers have discovered that using WaveShrink with the non-decimated wavelet transform can lead to lower mean-square-error predictions and fewer artifacts (see section 11.1). The wavelet shrinkage estimators are being applied to many types of problems, such as classification and discriminant analysis, density estimation, and time series prediction (see section 1.2 for references).

6.2 Understanding Wavelet Shrinkage

In order to best use the **waveshrink** function, you need to understand how and why WaveShrink estimators work. Estimation by WaveShrink rests on three simple principles:

1. Signal features can be represented by just a few wavelet coefficients, making wavelets good building blocks.

2. Noise affects all wavelet coefficients.

3. By shrinking wavelet coefficients towards zero, noise can be removed while preserving features.

6.2.1 Wavelets are Good Building Blocks

To illustrate the first principle, use the "bumps" signal, which consists of a series of peaks which decay rapidly towards zero. Create the bumps signal and its DWT with:

```
> bumps <- make.signal("bumps", n=1024)
> bumps.dwt <- dwt(bumps)
> plot(bumps.dwt)
```

The result is displayed in figure 6.3. Almost all of the wavelet coefficients are nearly zero. Even though the bumps signal has $n = 1024$ observations, we can represent the signal with just a fraction of wavelet coefficients $m \ll n$. In fact, the top 100 coefficients contain 99% of the total energy of the bumps signal.

6.2.2 Noise Affects All Wavelet Coefficients

Now generate a noisy signal following the signal-plus-noise model

$$y_i = f_i + \epsilon_i .$$

You can do this with the argument **snr** to the **make.signal** function, which sets the signal-to-noise ratio (by default **snr** is set to infinity, which gives a pure signal). Create a noisy bumps signal as follows:

```
> noisy.bumps <- make.signal("bumps", snr=3)
```

Next compute and plot the DWT of the **noisy.bumps** signal:

```
> plot(dwt(noisy.bumps))
```

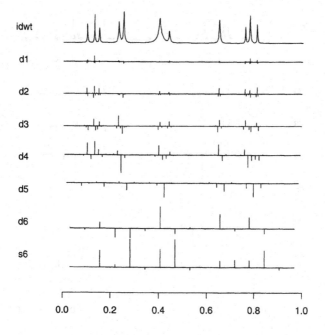

FIGURE 6.3. The bumps signal and its DWT.

The result is displayed in figure 6.4. In contrast to the DWT of the pure bumps signal, the wavelet coefficients are very noisy. This illustrates the second principle.

To understand why the second principle holds, decompose the wavelet transform $\mathbf{w} = \mathbf{W}\mathbf{y}$ of the noisy signal $\mathbf{y} = (y_1, y_2, \ldots, y_n)'$ as follows:

$$\mathbf{w} = \mathbf{W}\mathbf{f} + \mathbf{W}\epsilon$$

Since $\mathbf{W}$ is an orthonormal transform, $\mathbf{W}\epsilon$ is independent noise with a normal distribution and variance σ^2. Hence, every discrete wavelet coefficient w_i contributes noise of variance σ^2.

6.2.3 De-Noising by Wavelet Shrinkage

The heuristic for the WaveShrink procedure follows from the first two principles: by shrinking wavelet coefficients towards zero, we can eliminate noise while recovering the main features of the underlying signal or image. Apply the `waveshrink` function to the `noisy.bumps` signal and plot the DWT of the estimated signal:

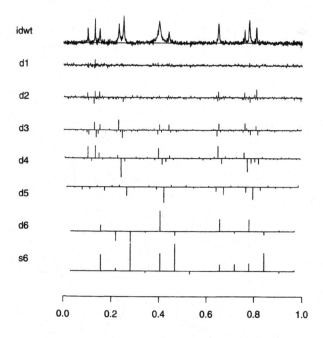

FIGURE 6.4. DWT of the `noisy.bumps` signal.

```
> smooth.bumps <- waveshrink(noisy.bumps)
> plot(dwt(smooth.bumps))
```

The result is displayed in figure 6.5. The DWT of the estimated signal is very similar to the DWT of the pure bumps signal. The main difference is that some fine scale coefficients $d_{1,k}$ and $d_{2,k}$ have been shrunk to zero. This accounts for the distortion in the estimate at the beginning of the signal.

6.2.4   The WaveShrink Algorithm

The WaveShrink algorithm is illustrated in figure 6.6. The three steps in the algorithm are:

[1] Apply the wavelet transform with J levels to the signal $\mathbf{Y}$, obtaining wavelet detail and smooth coefficients $\mathbf{d}_1, \mathbf{d}_2, \ldots, \mathbf{d}_J$, $\mathbf{s}_J$.

[2] Shrink the detail coefficients at the j finest scales to obtain new detail coefficients $\tilde{\mathbf{d}}_1 = \delta_{\lambda_1 \sigma_1}(\mathbf{d}_1), \ldots, \tilde{\mathbf{d}}_j = \delta_{\lambda_j \sigma_j}(\mathbf{d}_j)$. The

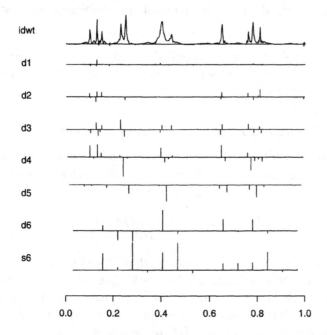

FIGURE 6.5. DWT of the smooth.bumps signal.

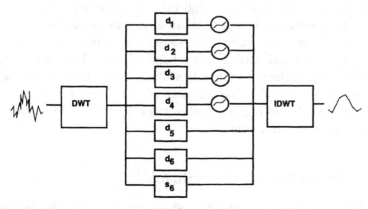

FIGURE 6.6. The WaveShrink algorithm for de-noising signals and smoothing densities.

function $\delta_c(x)$ shrinks x towards zero and is parameterized by $c = \lambda\sigma$ where λ is the threshold and σ is an estimate of the scale of the noise.

[3] Apply the inverse wavelet transform using the detail coefficients

$\tilde{\mathbf{d}}_1, \ldots, \tilde{\mathbf{d}}_j, \mathbf{d}_{j+1}, \ldots, \mathbf{d}_J, \mathbf{s}_J$ to obtain the WaveShrink estimate $\hat{\mathbf{f}}$.

In figure 6.6, shrinkage is applied to only the $j = 4$ finest levels. Note that both the threshold λ_i and noise scale factor σ_i may vary from level to level.

6.2.5   Theoretical Properties of Wavelet Shrinkage

Theoretical results show that for certain choices of the λ_j, the estimate $\hat{f}_{ws}$ provided by WaveShrink can almost achieve the *minimax risk* over a broad class of functions $\mathcal{F}$:

$$R(\hat{f}_{ws}, f) \approx \inf_{\hat{f}} \sup_{f \in \mathcal{F}} R(\hat{f}, f) \tag{6.1}$$

The optimality result (6.1) means that WaveShrink gives nearly the best possible estimate of $f(t)$ making a minimum of assumptions about the underlying nature of $\mathcal{F}$. For example, most nonparametric procedures assume $\mathcal{F}$ contains only smooth functions, but WaveShrink does not require $f(t)$ to be smooth.

The WaveShrink algorithm has a *locally adaptive bandwidth*. This means that at discontinuities and other non-smooth features, the WaveShrink estimate is more localized and uses a smaller band of data. For the smooth part of the signal, the WaveShrink estimate is less localized and uses a wider band of data. WaveShrink has been shown to perform remarkably well on a broad range of spatially inhomogeneous signals. The estimate is completely automatic; no tuning constants are required, other than the choice of the wavelet filter and thresholding rule.

6.3   Assessing the WaveShrink Fit

The first step in assessing the WaveShrink fit is to apply the summary function. Using the bumps example:

```
> summary(smooth.bumps)
Thresholds for WaveShrink and Noise Scale Estimates:
               s6    d6    d5    d4    d3    d2    d1
Thresholds 0.000 2.226 2.226 2.226 2.226 2.226 2.226
    Scales 0.215 0.215 0.215 0.215 0.215 0.215 0.215
```

```
Number of Non-zero Wavelet Coefficients:
       s6 d6 d5 d4   d3   d2   d1 total
signal 16  9 11 18   21   28   23   126
  data 16 16 32 64  128  256  512  1024
```

```
Statistics for Residuals: (897 degree of freedom)
    Min     1Q Median    3Q   Max Mean    SD   Mad
 -0.723 -0.158 -0.016 0.147 1.136    0 0.245 0.227
```

```
Normality Test for Residuals (Shapiro-Wilk Test)
W-statistic P-value
     0.97        0
```

```
Whiteness Test for Residuals (Box-Ljung Test)
Q-statistic P-value
   124.224        0
```

By default, the threshold is zero at the 2 coarsest resolution levels and shrinkage is applied only to the finest $J - 2 = 4$ levels. At those levels, a constant "universal" threshold of $\lambda_j = \sqrt{2 \log(n)} \approx 3.723$ is used. A single value $\sigma_j = 0.237$ is used to estimate the scale of the noise, and is obtained from the fine scale coefficients d_1.

The WaveShrink fit is based on a sum of 105 wavelet functions and the residuals have 918 degrees of freedom. The residuals are highly non-normal as indicated by the Shapiro-Wilk statistic [SM65]. The residuals are also highly autocorrelated as indicated by the Box-Ljung Q-statistic [LB78].

Use eda.plot to look at summary plots of the fit and the residuals. Start by looking at a standard EDA plot:

```
> eda.plot(smooth.bumps)
```

This results in figure 6.7.   The EDA plot for a waveshrink object produces four visual summaries:

Stack Plot	The stack.plot function decomposes the data into a signal (the WaveShrink fit) and a residual.
Box plot	The boxplot function creates box plots of the DWT coefficients for the original data with the WaveShrink thresholds superimposed as horizontal lines. Any wavelet coefficient lying inside the lines is set to zero.

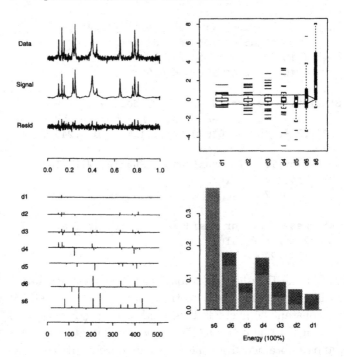

FIGURE 6.7. EDA Plot for WaveShrink estimate of `smooth.bumps`.

DWT Plot   The DWT of the signal is plotted.

Bar plot   For each crystal, the `barplot` function decomposes the energy of the data into the energy attributable to signal and residual energy. Energy for the data and signal is defined by

$$E_j^{\text{data}} \;=\; \sum_{k=1}^{n/2^j} d_{j,k}^2$$

$$E_j^{\text{signal}} \;=\; \sum_{k=1}^{n/2^j} \widetilde{d}_{j,k}^2$$

where $d_{j,k}^2$ and $\widetilde{d}_{j,k}^2$ are the original and shrunken wavelet coefficients respectively. The residual energy is defined as the difference $E_j^{\text{residual}} = E_j^{\text{data}} - E_j^{\text{signal}}$.

The EDA plot indicates that peaks have been oversmoothed. Get a closer look at this by looking at a "residual" EDA plot such as figure 6.8, which is produced as follows:

```
> eda.plot(smooth.bumps, resid=T)
```

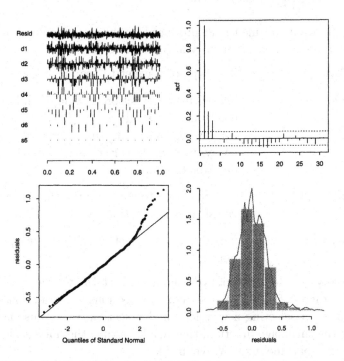

FIGURE 6.8. Residual EDA plot for WaveShrink estimate smooth.bumps. Top left: the DWT of the residual component. Top right: the autocorrelation function (acf) of the residual component. Bottom left: the quantile-quantile plot of the residual component versus the quantiles of a standard normal. Bottom right: a histogram and density estimate of the residuals.

The residual EDA plot consists of four diagnostic plots: the DWT of the residual component (top left), the autocorrelation function (acf) of the residual component (top right), the quantile-quantile plot of the residual component versus the quantiles of a standard normal (bottom left), and a histogram and density estimate of the residuals (bottom right). Because the peaks are oversmoothed, the distribution of the residual component is skewed toward high values. The oversmoothing also leads to significant autocorrelation in the residuals.

In the next section, you will learn how to tune the WaveShrink smooth by adjusting some parameters.

6.4   Tuning the Smooth

You may want to tune the fit produced by the **waveshrink** function. You can adjust any of four key arguments to change the fit:

wavelet	The type of wavelet to use for analysis.
shrink.fun	The shrinkage function $\delta_c(x)$ for shrinking the coefficients towards zero. (Recall that $c = \lambda_j\sigma_j$, where λ_j and σ_j are controlled, respectively, by the **shrink.rule** and **scale.rule** described below.)
shrink.rule	The rule used to compute the threshold λ_j in the shrinkage function.
scale.rule	The rule to use to estimate the scale σ_j of the noise.

As in any wavelet analysis, you should choose the wavelet to match the characteristics of your signal. See sections 5.1.2 and 12.4 for guidance on selecting a wavelet. In this section, we discuss how to select the shrinkage function, the shrinkage rule, the noise scale rule, and other parameters of WaveShrink.

Note: In later chapters, you will learn about some alternatives to the orthogonal DWT, including a non-decimated wavelet transform and a robust outlier-resistant wavelet transform (chapter 11), a wavelet packet transform (chapter 7), and a cosine packet transform (chapter 8). You can use the **waveshrink** function on these other types of transforms. You can learn about de-noising based on the non-decimated wavelet transform in section 11.1.1 and de-noising based on the robust wavelet transform in section 11.2.2. De-noising using the wavelet packet and the cosine packet transforms can also be done by the **waveshrink** function.

6.4.1   Selecting the Shrinkage Function

The argument **shrink.fun** is used to set the shrinkage function. Two basic forms for the shrinkage function δ_λ are the **soft** shrinkage and hard shrinkage defined by

$$\delta_\lambda^S(x) = \begin{cases} 0 & \text{if } |x| \le \lambda \\ \text{sign}(x)(|x| - \lambda) & \text{if } |x| > c \end{cases} \quad (6.2)$$

$$\delta_\lambda^H(x) = \begin{cases} 0 & \text{if } |x| \le \lambda \\ x & \text{if } |x| > \lambda \end{cases} \quad (6.3)$$

The shrinkage functions are implemented by the S+WAVELETS function shrink. To compare hard and soft shrinkage, shrink a linear function as follows to produce figure 6.9:

```
> par(mfrow=c(1,2))
> x <- rts(seq(from=-6,to=6,by=.02),start=-6, deltat=.02)
> hard.x <- shrink(x, threshold=3, shrink.fun="hard")
> soft.x <- shrink(x, threshold=3, shrink.fun="soft")
> plot(hard.x, type="l", ylim=c(-6,6), xlab="",
+       ylab="Hard Shrinkage")
> lines(x, lty=2)
> plot(soft.x, type="l", ylim=c(-6,6), xlab="",
+       ylab="Soft Shrinkage")
> lines(x, lty=2)
```

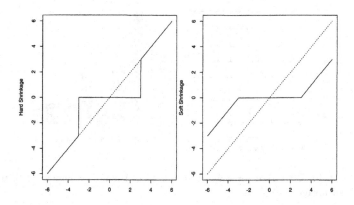

FIGURE 6.9. The shrinkage function with threshold $c = 3$ (solid line) applied to a linear function (dashed line). Left: "hard" shrinkage. Right: "soft" shrinkage.

The hard shrinkage function is so named because there is a discontinuity in the shrinkage function; values x which are above the threshold λ are untouched. By contrast, the soft shrinkage function is continuous since it shrinks values above the threshold λ.

By default, the soft shrinkage function is used. The motivation for soft shrinkage comes from the principle that the noise affects all wavelet coefficients (see section 6.2). Also, the continuity of the soft shrinkage function makes it preferable for statistical reasons.

In some applications, hard shrinkage may be preferred. This is particularly true when you are concerned about reducing the bias in estimating features. As an example, we smooth the bumps signal using the hard shrinkage function:

```
> eda.plot(waveshrink(noisy.bumps, shrink.fun="hard"))
```

The resulting EDA plot is shown in figure 6.10.   Hard shrinkage

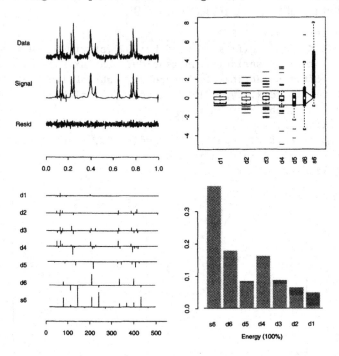

FIGURE 6.10. EDA Plot for wavelet shrinkage of the noisy.bumps signal using "hard" shrinkage.

does a better job in this example of preserving the bumps; compare with the EDA plot in figure 6.7 which uses soft shrinkage.

A comprehensive comparison between the soft and the hard shrinkage rules can be found in Bruce and Gao [BG95a]. Soft shrinkage suffers from very large bias in some examples. Hard shrinkage has smaller bias but has bigger variance. Also, hard shrinkage is very sensitive to small changes in the data. To remedy the drawbacks of hard shrink and soft shrink, you can use the semisoft shrinkage function $\delta_{\lambda_1,\lambda_2}$:

$$\delta_{\lambda_1,\lambda_2}(x) = \begin{cases} 0 & \text{if } |x| \le \lambda_1 \\ \text{sgn}(x)\frac{\lambda_2(|x|-\lambda_1)}{\lambda_2-\lambda_1} & \text{if } \lambda_1 < |x| \le \lambda_2 \\ x & \text{if } |x| > \lambda_2 \end{cases} \tag{6.4}$$

For values x near the lower threshold λ_1, $\delta_{\lambda_1,\lambda_2}(x)$ behaves like $\delta_{\lambda_1}^S(x)$. For values x above the upper threshold λ_2, $\delta_{\lambda_1,\lambda_2}(x) = \delta_{\lambda_2}^H(x)$ $= x$. Note that hard shrinkage, with $\lambda_1 = \lambda_2$, and soft shrinkage, with $\lambda_2 = \infty$, are limiting cases of (6.4). See Bruce and Gao [BG95b] for more details about the semisoft shrinkage function.

6.4.2 Selecting the Shrinkage Threshold

There are several different rules for selecting the threshold λ_j in the shrinkage function. Some of these rules are motivated from statistical theory, and offer a range of ways to shrink the coefficients. You specify a rule using the **shrink.rule** argument to **waveshrink**. The following rules for computing the thresholds λ_j are available in S+WAVELETS:

universal   The universal threshold is defined by $\lambda_j = \sqrt{2\log(n)}$ where n is the sample size. The universal threshold is the default method and yields the largest thresholds. As a result, using the universal threshold results in a relatively high degree of smoothness.

minimax   The minimax thresholds λ_j for soft and hard shrinkage have been computed by Donoho and Johnstone [DJ94] and Bruce and Gao [BG95a] for a range of sample sizes. The minimax threshold minimizes a theoretical upper bound on the asymptotic risk. They are always smaller than the universal threshold for a given sample size, and thus result in less smoothing.

adapt   Donoho and Johnstone [DJ95] propose a threshold which adapts to the signal at each multiresolution level. The threshold is based on the principle of minimizing the Stein Unbiased Risk Estimator (SURE) at each resolution level. The **adapt** threshold for crystal $\mathbf{d}_j$ with K coefficients is defined by

$$\lambda_j = \mathrm{argmin}_{t \geq 0} \mathrm{SURE}(\mathbf{d}_j, t)$$

where

$$\mathrm{SURE}(\mathbf{d}_j, t) = K - 2 \sum_{k=1}^{K} 1_{[|d_{j,k}| \leq t\sigma_j]}$$
$$+ \sum_{k=1}^{K} \min \left\{ (d_{j,k}/\sigma_j)^2, t^2 \right\}. \quad (6.5)$$

The **adapt** threshold can perform poorly if the coefficients are very sparse (i.e., most of the coefficients at a level are nearly zero). A test for sparseness is given by

$$\frac{1}{K} \sum_{k=1}^{K} \frac{d_{j,k}}{\sigma_j} - 1 \leq \frac{(\log_2 K)^{3/2}}{\sqrt{K}}. \quad (6.6)$$

If (6.6) holds for a vector of coefficients, then the universal threshold is used; otherwise the **adapt** threshold is used. This rule should be used only with the soft shrinkage.

top

Rather than choosing a threshold value based on statistical theory (as with the **universal**, **minimax**, and **adapt** thresholds), you can select the threshold by setting the number of coefficients which should be above the threshold. You do this in **waveshrink** by specifying **shrink.rule="top"** and using the argument **n.top** to specify the number of coefficients. This rule is equivalent to using the hard shrinkage with the $(1 - \mathbf{n.top}/n)$-th quantile of the absolute values of the wavelet coefficients as the threshold.

There is no hard and fast rule for which threshold to use—it depends on the application. The **universal** threshold gives you smoother estimates than `minimax` or `adapt`, but also has higher bias. The `top` threshold gives you the greatest control.

Here the adaptive threshold is used for the `noisy.bumps` example:

```
> eda.plot(waveshrink(noisy.bumps, shrink.rule="adapt"))
```

The resulting EDA plot is shown in figure 6.11. The adaptive thresh-

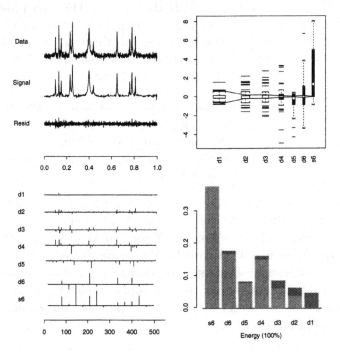

FIGURE 6.11. EDA Plot for wavelet shrinkage of the `noisy.bumps` signal using the adaptive threshold.

old does less smoothing of the peaks than the universal threshold; compare with figure 6.7. Note that the threshold varies from level to level. In particular, the threshold is relatively large at the finest scale, which is dominated by noise, and the threshold is relatively small at the coarsest scale, which is dominated by signal.

6.4.3 Estimating the Scale of the Noise

The shrinkage function δ_{c_j}, $c_j = \lambda_j \sigma_j$, depends on both the threshold rule λ_j, and the estimate σ_j of the scale of the noise. There are three rules for computing the scale factors σ_j:

d1   The finest scale detail coefficients $\mathbf{d}_1$ are used
                      to estimate a single scale factor for all levels:
                      $\sigma_j = \hat{\sigma}(\mathbf{d}_1)$.

all   All of the detail coefficients are used to es-
                      timate a single scale factor for all levels:
                      $\sigma_j = \hat{\sigma}(\mathbf{d}_1, \mathbf{d}_2, \ldots, \mathbf{d}_J)$. This can provide a
                      more accurate estimate of scale since it uses
                      all of the coefficients instead of just half of
                      them. However, it is also more likely to be
                      influenced by the signal.

each   A separate scale factor is estimated for each
                      crystal: $\sigma_j = \hat{\sigma}(\mathbf{d}_j)$ for $j = 1, 2, \ldots, J$. If the
                      scale of the noise varies from level to level,
                      then it is important to estimate a different
                      scale factor for each level.

The default method for selecting a scale factor is d1. This is the safest choice and sacrifices relatively little in efficiency in most cases.

You can also select the function $\hat{\sigma}$ used to compute the scale. By default, $\hat{\sigma}$ is the mad function, a highly robust estimate of scale. It is important to use a robust estimate of scale so that the signal does not leak into the estimate of the scale of the noise. You can provide your own scale estimator by using the argument scale.fun to the waveshrink function.

6.4.4   Choosing the Resolutions to Shrink

By default, shrinkage is applied to all J detail coefficients, where J is the maximum resolution level in the wavelet transform. You can restrict or increase the number of resolution levels to shrink using the argument smooth.levels.

In some applications, it is useful to set all the coefficients at fine scale resolution levels to zero. The argument zero.levels is used to specify which resolution levels are set to zero.

6.4.5   Extra-Fine Tuning

If the above tuning parameters are not enough, you can gain even greater control using the arguments vthresh and vscale. These arguments let you provide your own vector of thresholds and noise scales.

6.5   Comparison with Other Smoothers

Drawing on the extensive functionality of S-PLUS, you can compare the smooth produced by waveshrink with a variety of other methods. Some of the functions you can use in S-PLUS for smoothing include:

filter   Simple linear filters and moving averages.

median.smooth; smooth

> Simple robust smoothers based on running median filters.

ksmooth   Kernel smoothing.

supsmu   Nearest neighbor smoothing with locally adaptive bandwidth.

loess   Locally weighted regression smoothers.

smooth.spline   Spline smoothing.

tree   Piecewise constant approximations through classification and regression trees.

acm.filt; acm.smooth

> Approximate conditional mean (robust) Kalman filtering and smoothing.

For more information about these functions, refer to their respective online help files. S-PLUS for Windows users can find more information in volume 2 of the *S-PLUS User's Manual* and in the *S-PLUS for Windows Version 3.2 Supplement*. S-PLUS for UNIX users can find more information in the *S-PLUS Guide to Statistical and Mathematical Analysis*.

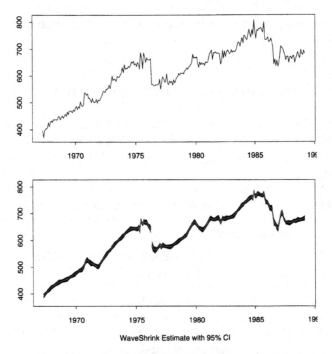

FIGURE 6.12. Seasonally adjusted monthly U.S. variety retail sales data (top plot) and WaveShrink estimate of the trend with approximate 95% confidence interval.

6.6   Variance and Bias Estimation in WaveShrink

Formulas for computing the exact pointwise bias and variance of WaveShrink have been derived by Bruce and Gao [BG95a]. The pointwise variances are computed by the function `var.waveshrink`. For orthogonal wavelets, `var.waveshrink` computes the variances using either a matrix formulation or a more efficient implementation that takes advantage of the sparsity in the matrix implied by the wavelet filters. For biorthogonal wavelets, which require considerably greater computational effort, only the more efficient implementation is used.

The mean, variance, L_2 risk, and covariances for the hard shrinkage and the soft shrinkage functions are computed by `wv.shrink.mean`, `wv.shrink.var`, `wv.shrink.l2`, and `wv.shrink.cov`.

The following example involves the problem of trend estimation for economic time series.

```
> # take out a linear trend
> census <- bvarrs.sa[3:266]
> census <- ts(census, start=1967+1/4, deltat=1/12)
> tt <- seq(from=1967.25, by=1/12, length=264)
> ls.trend <- lsfit(tt, census)
> census1 <- ls.trend$residuals
> lin.trend <- tt*ls.trend$coef[2]+ls.trend$coef[1]
> # Compute waveshrink estimate
> census2 <- waveshrink(census1, wavelet="s8",
+                boundary="periodic", n.levels=3)
> census.sd <- sqrt(var.waveshrink(census2,fast=F))
> par(mfrow=c(2,1))
> plot(census, type="l", xlab="", ylab="")
> plot(census2, type="n", ylab="", ylim=range(census),
+    xlab="WaveShrink Estimate with 95% CI")
> polygon(c(tt, rev(tt)),
+  c(census2+lin.trend-2*census.sd,
+    rev(census2+lin.trend+2*census.sd)),
+  border=F, col=1)
> lines(census2+lin.trend, col=0)
```

The result is displayed in figure 6.12. The top panel shows the seasonally adjusted monthly series of U. S. variety retail sales. The X-12-ARIMA [US 95] seasonal adjustment package was used to adjust this series. The second panel shows the WaveShrink estimate and approximate 95% confidence interval using hard shrinkage, the "s8" wavelet, and the universal threshold.

 Warning: To use the function var.waveshrink the boundary rule must be periodic.

6.7   Bootstrapping WaveShrink

This section explores the application of the bootstrap method to WaveShrink, another example of the type of analysis you can do in S+WAVELETS. The *bootstrap* is a fundamental tool to obtain variance estimates for procedures for which analytical results are either unattainable or unreliable [Efr79]. Since its introduction 15 years ago, the bootstrap has been used on a wide variety of estimation problems. The whitening property of the wavelet transform implies that the bootstrap should produce reasonable results in the context of WaveShrink and nonparametric regression estimation.

One approach to applying the bootstrap to WaveShrink is to re-sample from the wavelet coefficients of the residual in a nonpara-metric regression estimate. The basic idea is as follows: let $\mathbf{d}_1^r$, $\mathbf{d}_2^r$, $\mathbf{d}_3^r$, and $\mathbf{d}_4^r$ be the residuals in the wavelet domain after applying the threshold rule to the wavelet coefficients at the four finest res-olution levels. Form a new set of DWT coefficients by resampling from each of the resolution levels, either separately or together, and reconstruct from the new DWT to obtain a resampled series $\widehat{\mathbf{y}}^{(i)}$. Then apply wavelet shrinkage to $\widehat{\mathbf{y}}^{(i)}$ to obtain an estimated signal $\widehat{\mathbf{f}}^{(i)}$. A bootstrap estimate of scale can be obtained by repeating this procedure M times and computing a sample standard deviation from the estimates $\widehat{\mathbf{f}}^{(1)}$, $\widehat{\mathbf{f}}^{(2)}$, ..., $\widehat{\mathbf{f}}^{(M)}$.

To implement the bootstrap for the WaveShrink method, first de-fine the `resample.dwt` function:

```
> resample.dwt <- function(x, J=paste("d",1:4,sep=""))
+ {
+    for(j in J){
+       xj <- x[[j]]
+       x[[j]] <- sample(xj,replace=T)
+    }
+    reconstruct(x)
+ }
```

Now define the **bootstrap.ws** function based on the above proce-dure:

```
> bootstrap.ws <- function(y, M = 100, ...)
+ {
+    stats.fun <- function(x) quantile(x,p=c(.05,.95))
+    ws <- waveshrink(y, ...)
+    signal <- as.vector(ws)
+    resid.dwt <- attr(ws,"raw.dwt")-attr(ws,"clean.dwt")
+    samples <- matrix(0, length(signal), M)
+    for(i in 1:M){
+      new.resid <- resample.dwt(resid.dwt)
+      samples[,i] <- waveshrink(signal+new.resid, ...)
+    }
+    mat <- apply(samples,1,stats.fun)
+    attr.x <- attributes(ws)[c("class", "tsp")]
+    lower.ci <- mat[1,]
+    upper.ci <- mat[2,]
+    attributes(signal) <- attr.x
+    attributes(lower.ci) <- attr.x
```

```
+    attributes(upper.ci) <- attr.x
+    list(signal=signal, lower.ci=lower.ci,
+        upper.ci=upper.ci)
+ }
```

Since you are iterating the WaveShrink procedure 100 times, running this function will take a while. Try out the bootstrap function of a noisy blocks signal as follows:

```
> blocks <- make.signal("blocks",n=256,snr=6)
> blocks.ws <- bootstrap.ws(blocks, wavelet="haar",
+    shrink.fun="hard")
```

Finally, create the plot in figure 6.13 with

```
> par(mfrow=c(2,2))
> ylim <- range(unlist(blocks.ws), blocks)
> plot(make.signal("blocks", n=256), type="l", ylim=ylim,
+        ylab="noise-free signal")
> plot(blocks, type="l", ylim=ylim, ylab="noisy signal")
> plot(blocks.ws$signal, type="l", ylim=ylim,
+        ylab="WaveShrink estimate")
> plot(blocks.ws$signal, type="l", ylim=ylim,
+        ylab="90 % bootstrap confidence interval")
> x <- c(time(blocks),rev(as.vector(time(blocks))))
> y <- c(blocks.ws$lower,rev(as.vector(blocks.ws$upper)))
> polygon(x,y)
```

The variability of the estimate is much greater near the discontinuities. By contrast, the variability is very small in the middle of the piecewise constant blocks.

Warning: This example of the bootstrap procedure applied to WaveShrink is included for illustrative purposes only. Research is needed before such a procedure can be reliably used in practice. Procedures described in section 6.6 are much more reliable.

6.8 WaveShrink Applied to Spectral Density Estimation

Wavelet techniques have been applied to spectral density estimation problems by Gao [Gao93b, Gao93a], Moulin [Mou92, Mou93], Percival [Per93], Sachs and Schneider [vSS94] and Walden, McCoy and Percival [WMP95]. It can be shown that using wavelets to represent the spectral density leads to efficient estimates which are

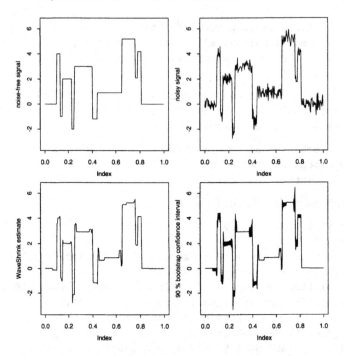

FIGURE 6.13. Bootstrapping with WaveShrink. Top left: the original noise-free blocks signal. Top right: the blocks signal plus random normal noise. Bottom left: the WaveShrink estimate of the signal. Bottom right: approximate 90% confidence pointwise intervals estimated using the bootstrap.

nearly as good as the optimal rate. The technique implemented in the S+WAVELETS toolkit was developed by Hong-Ye Gao [Gao93a]. This procedure, which falls in the general category of WaveShrink estimates for noisy data, is based on wavelet smoothing of the raw log-periodogram.

The estimation procedure applies the WaveShrink algorithm as described in section 6.2.4 to the log periodogram $g(\omega_k) = \log |X(\omega_k)|$ where $X(\omega_k)$ is the discrete Fourier transform and $\omega_k = 2\pi k$ are the fundamental frequencies. You need to use a special threshold rule to estimate the spectral density function because the noise model is non-Gaussian (the "data" is the log-periodogram and the "noise" has a Chi-square distribution if the time series is Gaussian). The threshold for level j is chosen according to the formula

$$\lambda_j = \max\left(\pi\sqrt{\log_e(n)/3}, \ \log_e(2n)2^{-(j-1)/4}\right)$$

where n is the length of the original series and j is the wavelet level (with 1 as the finest level and J the coarsest level). See [Gao93a] for details.

You can compute a WaveShrink estimate of the spectral density function using the function spec.wave. As an example, let us create a simulated autoregressive (AR) time series:

```
> arcoef <- c(-2.5216281, 4.7715359, -7.9199915,
+            11.9769211, -16.0778828, 20.6343346,
+            -25.0531521, 28.8738136, -31.8046265,
+            34.0071373, -34.7700272, 34.3151321,
+            -32.7861099, 30.2861233, -26.7109356,
+            22.8838310, -18.7432098, 14.5717688,
+            -10.7177744, 7.5322194, -4.7226319,
+            2.6807923, -1.3391306, 0.5167125)
> x <- rts(arima.sim(model=list(ar=-arcoef), n=2048),
+    start=0, deltat=1/2048)
> ts.plot(x, type="l", xlab="AR time series", ylab="")
```

The simulated time series is shown in figure 6.14.

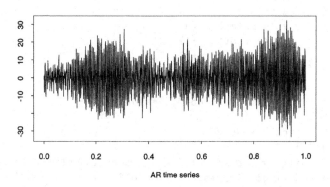

FIGURE 6.14. A simulated AR series which has sharp peaks in its spectral density function.

These AR coefficients correspond to an AR process with complex roots which have the following modulus and phase

$$
\text{mod} =
\begin{pmatrix}
1.03 \\
1.03 \\
1.005 \\
1.03 \\
1.03 \\
1.03 \\
1.03 \\
1.03 \\
1.03 \\
1.03 \\
1.03
\end{pmatrix}
\qquad
\text{arg} =
\begin{pmatrix}
0.2 \\
0.4 \\
0.6 \\
0.8 \\
1.0 \\
1.4 \\
1.6 \\
1.8 \\
2.0 \\
2.2 \\
2.4 \\
2.6
\end{pmatrix}
$$

The true spectrum for this AR series has sharp peaks. To compute the true spectrum for an autoregressive moving-average (ARMA) process, define the following function:

```
> true.arma.spec <- function(model, log=T)
+ {
+   spectra <- function(coef, J=8){
+     p <- length(coef)
+     n <- 2^J
+     x <- pi*c(0:n)/n
+     cc <- rep(1, n+1)
+     ss <- rep(0, n+1)
+     if(p>0) for(i in 1:p){
+       cc <- cc - coef[i]*cos(i*x)
+       ss <- ss - coef[i]*sin(i*x)
+     }
+     cc^2 + ss^2
+   }
+   if(log) sp <- 10*(log10(spectra(model$ma))-
+     log10(2*pi*spectra(model$ar)))
+   else sp <- spectra(model$ma)/(2*pi*spectra(model$ar))
+   n <- length(sp)-1
+   x <- pi*c(0:n)/n
+   invisible(list(freq=x, spec=sp))
+ }
```

Now compute the true spectrum, along with the log-periodogram, the smoothed log-periodogram using a triangular spectral window, and the WaveShrink estimate for the simulated series:

```
> true <- true.arma.spec(list(ar=-arcoef))
> period.est <- spec.pgram(x, plot=F)
> tri.est <- spec.pgram(x,spans=c(21,21), plot=F)
> wave.est <- spec.wave(x, wavelet="bs1.5", n.levels=5,
+          shrink.level=4, plot=F)
```

The triangular spectral window is obtained by setting the argument
spans=c(21,21), which applies two passes of a rectangular window
of width 21. The width of the triangular window was choosen so
that the estimated spectrum is roughly as smooth as the WaveShrink
estimate, with respect to the mean absolute first differences.

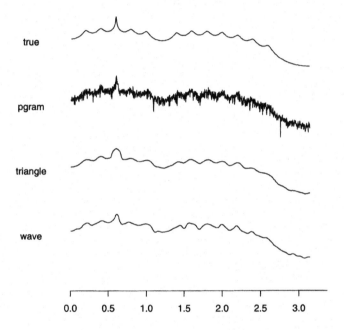

FIGURE 6.15. The true spectrum for a simulated AR series (top), the raw
log-periodogram (second), the smoothed log-periodogram using a triangular spectral window (third), and the WaveShrink estimate (bottom).

Save the spectra as a list and use **stack.plot** to compare the
estimates:

```
> spec.est <- list(true=true$spec[-1],
+    pgram=period.est$spec[-1],
+    triangle=tri.est$spec[-1],
+    wave=wave.est$spec[-1])
> stack.plot(spec.est, times=true$freq[-1], same.scale=T)
```

This results in figure 6.15. The traditional nonparametric estimator based on a triangular spectral smoothing window tends to over-smooth the peaks. Using shorter span smoothing windows would better preserve the peaks, but would result in an unnecessarily rough estimate elsewhere. The WaveShrink estimator is equivalent to using a variable bandwidth smoother. It preserves the peaks while producing a smooth estimate elsewhere.

 Warning: The function `spec.wave` behaves like the other spectral density estimation functions `spec.pgram` and `spec.ar`, and not like the `waveshrink` function. For more information about these functions, see their respective online help files.

7

Wavelet Packet Analysis

This chapter discusses *wavelet packet* analysis for one-dimensional signals. You will learn how to do following:

- Create and plot wavelet packet functions using the function `wavelet.packet` (see section 7.1).

- Create a wavelet packet table, which encompasses the whole range of wavelet packet transforms (WPT's), using the function `wp.table` (see section 7.2).

- Select orthogonal transforms, such as the DWT, from a wavelet packet table (see section 7.3).

- Select optimal transforms using the Coifman and Wickerhauser "best basis" algorithm with the `best.basis` function (see section 7.4).

- Perform a wavelet packet analysis of a signal (see section 7.5).

7.1 Wavelet Packets

Wavelet packet analysis is an important generalization of wavelet analysis, pioneered by Ronald Coifman, Yves Meyer, Victor Wickerhauser, and other researchers [CMQW90, CM91, CW92, CMW92,

Wic94a]. Wavelet packet functions comprise a rich family of building block functions. Wavelet packet functions are still localized in time, but offer more flexibility than wavelets in representing different types of signals. In particular, wavelet packets are better at representing signals that exhibit oscillatory or periodic behavior.

Wavelet packet functions are generated by scaling and translating a family of basic function shapes, which include father wavelets $\phi(t)$ and mother wavelets $\psi(t)$. In addition to $\phi(t)$ and $\psi(t)$, there is a whole range of wavelet packet functions $W_b(t)$. These functions are parameterized by an *oscillation* or frequency index b. A father wavelet corresponds to $b = 0$, so $\phi(t) \equiv W_0(t)$. A mother wavelet corresponds to $b = 1$, so $\psi(t) \equiv W_1(t)$. Larger values of b correspond to wavelet packets with more oscillations and higher frequency. See section 13.1 to learn how to derive wavelet packet functions.

To create a wavelet packet, use the function `wavelet.packet`. The argument `oscillation=b` specifies the oscillation index b. Create the wavelet packet in figure 7.1 as follows:

```
> plot(wavelet.packet(oscillation=5))
```

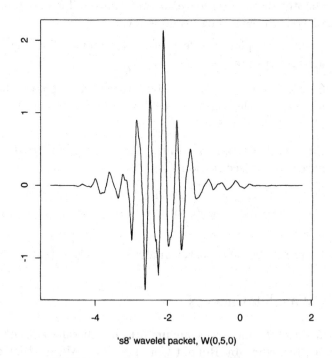

's8' wavelet packet, W(0,5,0)

FIGURE 7.1. A wavelet packet function.

7.1.1   Wavelet Packet Families

As in wavelet analysis, there are different families of wavelet packets. In fact, you can use the same families as for wavelets—table 2.1 on page 17 lists the orthogonal wavelets and table 2.2 on page 35 lists the biorthogonal wavelets. For example, create the family of **haar** wavelet packets $W_0(t)$, $W_1(t)$, ..., $W_7(t)$ as follows to produce figure 7.2.

```
> par(mfrow=c(4,2))
> for(b in 0:7)
+    plot(wavelet.packet("haar", oscillation=b))
```

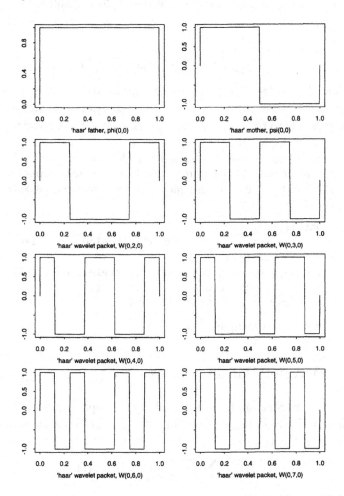

FIGURE 7.2. The family of **haar** wavelet packets $W_0(t)$, $W_1(t)$, ..., $W_7(t)$.

For the **haar** wavelet packet, the index b corresponds to the number of "zero crossings." The wavelet packet $W_0(t)$ has no zero crossings while the wavelet packet $W_7(t)$ has seven zero crossings. For other types of wavelet packets, like **s8**, the number of oscillations increases with b, although the number of zero crossings is typically greater than b.

7.1.2   Wavelet Packet Location and Scale Families

Wavelet packet approximations are based on translated and scaled wavelet packet functions $W_{j,b,k}$. These are generated from the base functions W_b as follows:

$$W_{j,b,k}(t) = 2^{-j/2}W_b(2^{-j}t - k). \tag{7.1}$$

The wavelet packet $W_{j,b,k}$ has scale 2^j and location $2^j k$. As in wavelet analysis, the index j corresponds to the resolution level and the index k corresponds to the translation shift. However, for a wavelet packet, there is also the oscillation parameter b.

To create scaled and translated wavelet packets $W_{j,b,k}$, use the optional arguments **level** and **shift** to the **wavelet.packet** function. For example, create and plot the **haar** wavelet packets $W_{0,4,2}$ and $W_{2,4,0}$ as follows:

```
> par(mfrow=c(1,2))
> wp1 <- wavelet.packet("haar", level=0, oscill=4,
+         shift=2)
> wp2 <- wavelet.packet("haar", level=2, oscill=4)
> plot(wp1, xlim=c(0,4), ylim=c(-1, 1))
> plot(wp2, xlim=c(0,4), ylim=c(-1, 1))
```

These are plotted in figure 7.3.

7.2   Wavelet Packet Tables

In wavelet packet analysis, a signal $f(t)$ is represented as a sum of orthogonal wavelet packet functions $W_{j,b,k}(t)$ at different oscillations, scales, and locations:

$$f(t) \approx \sum_j \sum_b \sum_k w_{j,b,k} W_{j,b,k}(t). \tag{7.2}$$

The range of the summation for the levels j and the oscillations b is chosen so that the wavelet packet functions are orthogonal.

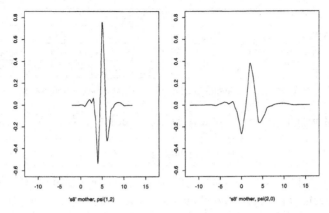

FIGURE 7.3. Scaled and translated **haar** wavelet packets $W_{0,4,2}$ and $W_{2,4,0}$.

The theoretical wavelet packet coefficients are given by the integral

$$w_{j,b,k} \approx \int W_{j,b,k}(t)f(t)dt. \qquad (7.3)$$

In wavelet analysis, given a maximum resolution level J, the wavelet functions used to represent a signal are fixed. By contrast, in wavelet packet analysis, one chooses many different combinations of wavelet packet functions in (7.2) to represent a signal. This leads to a whole set of possible wavelet packet transformations.

Wavelet packet analysis starts with construction of a *wavelet packet table*. Like the DWT, a wavelet packet table has coefficients at different resolution levels and translations. However, a wavelet packet table also has coefficients corresponding to different oscillations. At resolution level j, the table has wavelet packet coefficients with oscillation indices $b = 0, 1, \ldots, 2^j - 1$. By contrast, for each resolution level, the DWT has coefficients at just one oscillation index (or two at the coarsest level).

Suppose you have n sampled signal values $\mathbf{f} = (f_1, f_2, \ldots, f_n)'$ where n is a multiple of 2^J. The wavelet packet table has $J + 1$ resolution levels where J is the maximum resolution level. At resolution level j, a table has n coefficients, divided into 2^j *coefficient blocks*, or *crystals*, to use the terminology of chapter 2 (see page 23). When you stack the $J + 1$ resolution levels on top of one another, you get the $(J + 1) \times n$ table of coefficients, divided into 2^{J+1} crystals. A crystal is a set of coefficients arranged on a lattice. A wavelet packet

Level 0	$\mathbf{w}_{0,0}$							
Level 1	$\mathbf{w}_{1,0}$				$\mathbf{w}_{1,1}$			
Level 2	$\mathbf{w}_{2,0}$		$\mathbf{w}_{2,1}$		$\mathbf{w}_{2,2}$		$\mathbf{w}_{2,3}$	
Level 3	$\mathbf{w}_{3,0}$	$\mathbf{w}_{3,1}$	$\mathbf{w}_{3,2}$	$\mathbf{w}_{3,3}$	$\mathbf{w}_{3,4}$	$\mathbf{w}_{3,5}$	$\mathbf{w}_{3,6}$	$\mathbf{w}_{3,7}$

TABLE 7.1. Wavelet packet table with 3 resolution levels.

crystal $\mathbf{w}_{j,b}$ is indexed by level j and oscillation b:

$$\mathbf{w}_{j,b} = (w_{j,b,1}, w_{j,b,2}, \cdots, w_{j,b,n/2^j})'.$$

Table 7.1 shows the layout of a wavelet packet table with 3 resolution levels. The level 0 coefficients in the table $\mathbf{w}_{0,0} = (w_{0,0,1}, w_{0,0,2}, \ldots, w_{0,0,n})$ are equal to the original signal: $w_{0,0,k} \equiv f_k$. The level 1 crystals $\mathbf{w}_{1,0}$ and $\mathbf{w}_{1,1}$ have scale 2 and correspond to the DWT crystals $\mathbf{s}_1$ and $\mathbf{d}_1$. There are four crystals at level 2: $\mathbf{w}_{2,0}$, $\mathbf{w}_{2,1}$, $\mathbf{w}_{2,2}$, and $\mathbf{w}_{2,3}$. In general, at resolution j, there are 2^j crystals with oscillations $b = 0, 1, \ldots, 2^j - 1$

Note: In S+WAVELETS, wavelet packet tables are stored in "sequency" order, which corresponds to increasing oscillation index. See [Wic94a] for alternative ways to store wavelet packet tables.

Orthogonal wavelet packet approximations of the form (7.2) involve particular subsets of n coefficients from a wavelet packet table. In sections 7.3 and 7.4, you will learn how to select a wavelet packet transform (WPT) from a table. This section concentrates on the wavelet packet table as a whole.

Note: If n is not a multiple of 2^J, the number of coefficients in the (j, b)th block of a wavelet packet table may not be exactly $n/2^{j-1}$. However, there are still n coefficients at each level.

7.2.1 Computing a Wavelet Packet Table

In S+WAVELETS, you create a wavelet packet table with the function **wp.table**. Create and plot a wavelet packet table for a linear chirp with:

```
> lc <- make.signal("linchirp", n=1024)
> lc.tab <- wp.table(lc, boundary="zero")
> plot(lc.tab)
```

The plot is given in figure 7.4. Each resolution level has $n = 1024$ coefficients and is divided into blocks, indicated by dashed grid lines, according to the oscillation index (see table 7.1). Level 0 reproduces

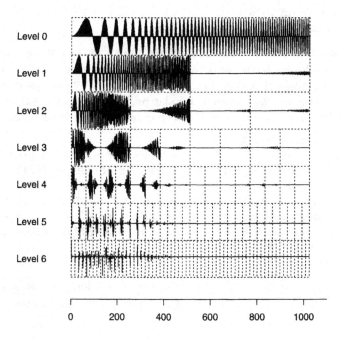

FIGURE 7.4. Wavelet packet table for the linear chirp signal.

the original linear chirp signal. Resolution levels 1 through 6 are plotted below. On each level, the leftmost block has oscillation index $b = 0$ and the rightmost block has oscillation index $b = 2^j - 1$. Hence, low frequency coefficients are on the left side of the table and high frequency coefficients are on the right.

The wavelet packet coefficient $w_{j,b,k}$ is plotted as a vertical line extending from zero. It is located at the kth coefficient in the bth block of the jth row. The coefficients in a given level are all plotted on the same vertical scale. The x axis shows the coefficient index for level 0.

7.2.2 Wavelet Packet Table Objects

Print out the wavelet packet table for the linear chirp as follows:

```
> lc.tab
Wavelet Packet Table for  lc
Wavelet: s8
Length of series: 1024
```

```
Number of levels: 6
Boundary correction rule: zero
```

By default, a wavelet packet table uses the s8 wavelet, a maximum of $J = 6$ resolution levels, and the **periodic** boundary correction rule. You can change these options just as for the DWT; see section 5.1.

Extracting crystals from a wavelet packet table is the same as for the DWT. For example, extract the crystal $\mathbf{w}_{6,4}$ from lc.tab with:

```
> lc.tab[["w6.4"]]
   w6.4(1)    w6.4(2)   w6.4(3)    w6.4(4)    w6.4(5)
0.02060631 -0.2215204 0.703054 -2.587479 -6.762177
   w6.4(6)    w6.4(7)   w6.4(8)   w6.4(9)    w6.4(10)
1.642022 -0.1486469 -0.1812057 0.1955425 -0.003135036
   w6.4(11)  w6.4(12)  w6.4(13)   w6.4(14)    w6.4(15)
-0.3956336 0.5876085 0.4023608 -0.00787465 -0.007013531
   w6.4(16)
-0.00198717
```

A wavelet packet table object is stored as a long vector of coefficients. The coefficients are ordered starting from the top row of the table.

7.3   Wavelet Packet Transforms

A wavelet packet table is a highly redundant representation of a signal: you start with n coefficients and end up with $(J + 1) \times n$ coefficients. To analyze the signal, it is often useful to select n coefficients from the table to create an orthogonal (linearly independent) *wavelet packet transform* (WPT) from the table. Only very special sets of coefficients give orthogonal transforms. One example of an orthogonal WPT is the DWT. There are many other wavelet packet transforms as well.

You can obtain the $(J + 1)n$ wavelet packet table coefficients $\widetilde{\mathbf{w}}$ through the linear transformation

$$\widetilde{\mathbf{w}} = \widetilde{\mathbf{W}}\mathbf{f}. \tag{7.4}$$

where $\widetilde{\mathbf{W}}$ is a $(J+1)n \times n$ matrix. A WPT is equivalent to a selection of n linearly independent rows from the matrix $\widetilde{\mathbf{W}}$ to obtain an orthogonal $n \times n$ matrix $\mathbf{W}$:

$$\mathbf{w} = \mathbf{W}\mathbf{f}. \tag{7.5}$$

To compute a wavelet packet table or the WPT, you don't actually perform the matrix multiplication (7.5). Instead, you use a fast algorithm similar to the pyramid algorithm; see section 13.1.

In this section, you will learn how to select orthogonal transforms from a wavelet packet table. In section 7.4, you will learn how to automatically select "optimal" transforms from a table.

7.3.1 Obtaining the DWT From a Table

The DWT can be obtained by selecting the crystals $\mathbf{w}_{1,1} = \mathbf{d}_1$, $\mathbf{w}_{2,1} = \mathbf{d}_2, \ldots, \mathbf{w}_{J,1} = \mathbf{d}_J$, and $\mathbf{w}_{J,0} = \mathbf{s}_J$ from the table. In S+WAVELETS this can be done with the function `as.dwt`. Create and plot a DWT object from the wavelet packet table `lc.tab` with:

```
> lc.dwt <- as.dwt(lc.tab)
> plot(lc.dwt)
```

The result is shown in figure 7.5. This is the same DWT that would

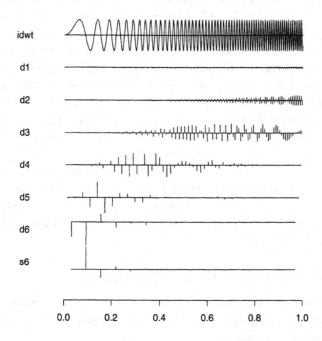

FIGURE 7.5. DWT of the linear chirp signal obtained from a wavelet packet table.

be produced using the `dwt` function applied to the linear chirp signal.

7.3.2 Selecting a Level

You don't go to the trouble of computing a wavelet packet table merely to obtain the DWT. Another important transform, which is widely used in engineering applications, is obtained by selecting an entire resolution level from a packet table. For example, a level 4 transform corresponds to the crystals $\mathbf{w}_{4,0}$, $\mathbf{w}_{4,1}$, ..., $\mathbf{w}_{4,2^4-1}$. Create a level 4 transform for the linear chirp by

```
> lc.4 <- lc.tab[level=4]
```

The level argument to the [subscript operator selects an entire resolution level from a table.

Create a stack plot of lc.4, shown in figure 7.6, with

```
> stack.plot(lc.4, bars=T)
```

All crystals in the level 4 wavelet packet transform (WPT) have $1024/2^4 = 64$ coefficients. Each crystal is plotted with its own vertical scale. The argument bars=T places vertical bars on the right hand side to compare the range of the crystals. The vertical bars are all the same absolute length, so a crystal with a very wide bar has a very small dynamic range.

Note: There are some large coefficients at the right boundary for the level 4 transform of the chirp signal. This is because you used the zero boundary rule in creating the wavelet packet table, and the signal is non-zero at the right boundary. If you use the default periodic boundary rule, you will see large coefficients at both boundaries. For the DWT, you can avoid boundary artifacts using the boundary rule poly0. For wavelet packet tables, however, the boundary rule poly0 will often lead to large boundary coefficients. To avoid the boundary artifacts for wavelet packet tables (and ensure perfect reconstruction), you may need to use a biorthogonal wavelet with the reflection boundary rule. See chapter 14 for details concerning boundary conditions.

7.3.3 General Wavelet Packet Transforms

An orthogonal transform can be created by selecting a set of crystals $\mathbf{w}_{j,b}$ from a wavelet packet table. The set of crystals must satisfy the following two conditions:

1. Every column in the wavelet packet table is covered by one crystal. This ensures the transform can be inverted to reconstruct the signal.

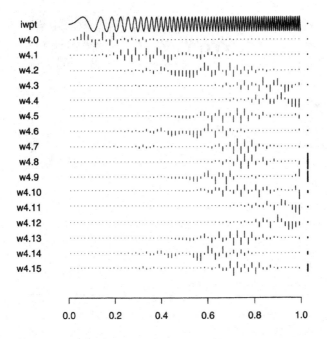

FIGURE 7.6. A level 4 packet transform for the linear chirp signal.

2. No column in the wavelet packet table has more than one crystal. This ensures that the transform is orthogonal.

You can select your own wavelet packet transform (WPT) using the subscript operator [. This works for wavelet packets in the same way as for DWT objects (see section 5.2). We select and plot the following basis from lc.tab:

```
> wp.nms <- c("w2.0", "w2.1", "w4.8", "w5.18", "w5.19",
+       "w3.5", "w3.6", "w4.14", "w6.60", "w6.61",
+       "w6.62", "w6.63")
> lc.mywpt <- lc.tab[wp.nms]
> stack.plot(lc.mywpt)
```

The transform is shown in figure 7.7.

The function **pgrid.plot** shows the positions of the crystals of a wavelet packet transform in a table. We can use **pgrid.plot** to compare the wavelet packet transforms computed for the linear chirp:

```
> par(mfrow=c(2,2))
> plot(lc.tab)
> pgrid.plot(lc.dwt)
```

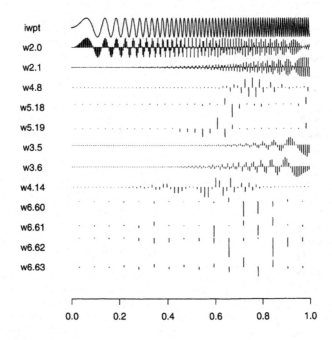

FIGURE 7.7. The wavelet packet transform `lc.mywpt`.

```
> pgrid.plot(lc.4)
> pgrid.plot(lc.mywpt)
```

The result is given in figure 7.8. All transforms satisfy conditions 1 and 2 for creating an orthogonal transform.

Note: You can directly compute a wavelet packet transform (WPT) from a signal with the function **wpt**. For example, to create the resolution 4 WPT, you can use the expression

```
lc.4 <- wpt(lc, n.level=4)
```

See the **wpt** online help file for details.

7.3.4 Inverting Wavelet Packet Transforms

You can recover the original signal from a wavelet packet transform (WPT) by applying the inverse wavelet packet transform (IWPT). The IWPT can be computed in S+WAVELETS with the function **reconstruct**. Invert the level 4 WPT of the linear chirp and compute the relative L_2 error of the reconstruction with:

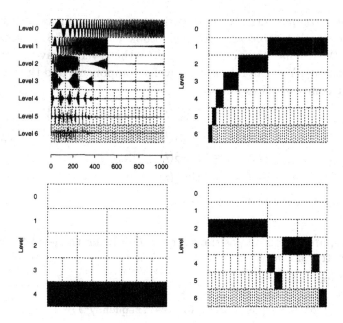

FIGURE 7.8. Comparison of wavelet packet transforms for a linear chirp. Top left: the wavelet packet table. Top right: the DWT in a wavelet packet table. Bottom left: the resolution level 4 transform. Bottom right: the transform lc.mywpt.

```
> lc.recon <- reconstruct(lc.4)
> vecnorm(lc - lc.recon)/vecnorm(lc)
[1] 8.124546e-13
```

Note: You can also use the function iwpt to compute the inverse wavelet packet transform. See the wpt online help file for details.

7.3.5 Decomposing a Wavelet Packet Transform

A wavelet packet transform can be used to decompose a signal into orthogonal components

$$W_{j,b}(t) = \sum_k w_{j,b,k} W_{j,b,k}(t). \tag{7.6}$$

This is analogous to the multiresolution decomposition discussed in section 2.3. For example, the level 4 wavelet packet decomposition is

$$f(t) \approx W_{4,0}(t) + W_{4,1}(t) + \cdots + W_{4,15}(t)$$

Signal components reconstructed from overlapping columns of a wavelet packet table are identical. Look at figure 7.8 to see how the DWT signal components relate to the level 4 wavelet packet components. You can relate the wavelet components and the wavelet packet components as follows:

$$\begin{aligned}
\mathbf{D}_1(t) &= W_{4,8}(t) + W_{4,9}(t) + \cdots + W_{4,15}(t) \\
\mathbf{D}_2(t) &= W_{4,4}(t) + W_{4,5}(t) + W_{4,6}(t) + W_{4,7}(t) \\
\mathbf{D}_3(t) &= W_{4,2}(t) + W_{4,3}(t) \\
\mathbf{D}_4(t) &= W_{4,1}(t) \\
\mathbf{S}_6(t) + \mathbf{D}_6(t) + \mathbf{D}_5(t) &= W_{4,0}(t).
\end{aligned}$$

Compare the decompositions for the DWT and the level 4 wavelet packet transform with:

```
> par(mfrow=c(1,2))
> plot(decompose(lc.dwt))
> plot(decompose(lc.4, order="frequency"))
```

The result is shown in figure 7.9.   The DWT gives a more refined

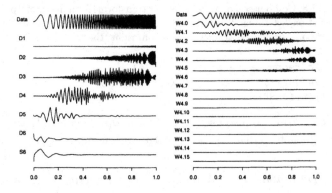

FIGURE 7.9. Comparison of wavelet decomposition with wavelet packet decomposition. Left: the wavelet decomposition. Right: level 4 wavelet packet decomposition.

decomposition of coarse features, but the wavelet packet transform gives a more refined decomposition of fine scale features. For the linear chirp, the wavelet packet transform offers advantages over the DWT. With a more refined decomposition of the fine scale features, wavelet packets are better at localizing the frequency behavior at high frequencies. The next section illustrates this using "time-frequency" plots.

7.3.6   Time-Frequency Plots for Wavelet Packets

Time-frequency analysis is concerned with how the frequency representation of the signal changes over time. Since the wavelet packet approximation (7.2) is localized in time and separates a signal into different *frequencies*, it can be used as a tool for time-frequency analysis [CMW92]. Using a wavelet packet transform, you can construct a *time-frequency plot* by dividing the time-frequency plane into rectangles. The modulus of each wavelet packet coefficient determines the color (or gray level) of each rectangle.

Time-frequency analysis is related to time-scale analysis, which was discussed in section 2.4. In fact, just as the DWT is a special case of wavelet packet transforms, the time-scale plot is a special case of the time-frequency plot. A time-frequency plot is also closely related to the spectogram, a tool commonly used to visualize the changing frequency behavior of a signal over time [RV91].

Note: Widely used techniques in time-frequency analysis are the *short-time Fourier transform* (STFT) and the *spectogram*. There are a variety of functions in S+WAVELETS for computing and using a local cosine transform, which is very closely related to the STFT (see chapter 8). Other methods for time-frequency analysis include quadratic (bilinear) time-frequency representations, such as the Wigner distribution. These provide better time-frequency concentration. Refer to [HBB92] for a recent review article on time-frequency analysis.

The function `time.freq.plot` is used to display a time-frequency plot. Compute the time-frequency plot for the linear chirp signal using a level 4 wavelet packet transform:

```
> time.freq.plot(lc.4)
> abline(a=0, b=.25*512, lwd=2)
```

The result is shown in figure 7.10.   The linear chirp should show up as a linear function in the time-frequency plane, as indicated in figure 7.10 by the line drawn by the `abline` function. You can see that the wavelet packet time-frequency plot does a reasonably good job of isolating the chirp.

As in a time-scale plot, each wavelet packet coefficient occupies a box having a constant area. The height of the box depends on the

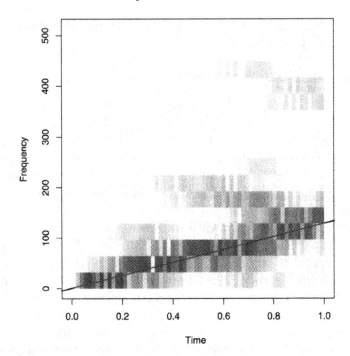

FIGURE 7.10. The time–frequency plot using a level 4 wavelet packet transform for a linear chirp.

scale for the wavelet packet: fine scale coefficients occupy tall thin boxes and coarse scale coefficients occupy flat wide boxes. Since all coefficients for a level 4 wavelet packet transform have the same scale (2^4), all boxes have the same size in figure 7.10.

By default, the color (or gray level) of each box corresponds to the square root of the absolute value of a cosine packet coefficient $w_{j,b,k}$. The horizontal and vertical center of a box is located roughly at the time and frequency *center* of the associated wavelet packet function. The width and height of the box is given by the time and frequency *bandwidth*. For a signal sampled at times t_0, $t_0 + \Delta_t$, $\ldots$, $t_0 + (n-1)\Delta_t$, the center of the box for coefficient $w_{j,b,k}$ is

$$(x, y) = (t_0 + (2^j(k - .5) + .5)\Delta_t, (b + .5)/2^j).$$

The width and height of the box are given by

$$(\Delta x, \Delta y) = (2^j \Delta_t, 1/2^j).$$

Compare the time–frequency plot to the time–scale plot computed from the DWT of the linear chirp with:

```
> time.freq.plot(lc.dwt)
> abline(a=0, b=.25*512, lwd=2)
```

The time-scale plot, shown in figure 7.11, does not localize the frequency behavior of the chirp as well as the wavelet packet time-frequency plot.

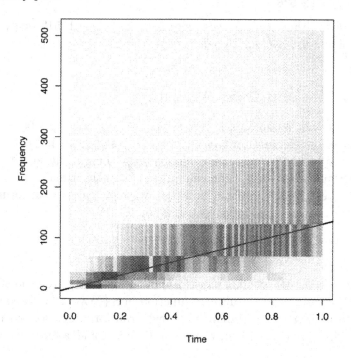

FIGURE 7.11. The time-scale plot for a linear chirp.

In time-frequency plots, the tradeoff between time and frequency is analogous to the tradeoff between time and inverse scale:

time bandwidth × frequency bandwidth ≥ constant.

This is known as the "Heisenberg uncertainty principle."

Warning: If the number of sample values n is not divisible by 2^J where J is the number of levels in the transform, then a time-frequency plot cannot be produced. For this reason it is often a good idea to restrict yourself to sample sizes divisible by 2^J in wavelet packet analysis.

Use the function **nice.n** to find the biggest integer not greater than n that is divisible by 2^J and the smallest integer not less than

n that is divisible by 2^J. For example, suppose your signal length is $n = 534$ and you want to obtain a wavelet packet decomposition with $J = 6$ levels. Use `nice.n` to obtain the next lower and next larger acceptable signal lengths:

```
> nice.n(n=534, J=6)
[1] 512 576
```

You must either truncate your signal to have length 512 or extend your signal to have length 576.

7.4   The Best Basis Algorithm

Coifman and Wickerhauser [CW92] developed the "best basis" algorithm for selecting optimal bases (i.e., transforms) from wavelet packet tables. The best basis algorithm automatically adapts the transform to best match the characteristics of the signal.

The best basis algorithm finds the wavelet packet transform $\mathcal{W}$ that minimizes an additive cost function E:

$$E(\mathcal{W}) = \sum_{j,b \in \mathcal{I}} E(\mathbf{w}_{j,b}) \tag{7.7}$$

where $\mathcal{I}$ is the set of index pairs (j, b) of the crystals in the transform $\mathcal{W}$. Minimizing the default cost function in S+WAVELETS is equivalent to finding the minimum "entropy" transform. See section 7.4.4 for a discussion of the entropy cost function and alternative cost functions.

7.4.1   Computing a Best Basis

Use the `best.basis` function to compute the best basis from a wavelet packet table. We try this for the linear chirp signal with:

```
> lc.bb <- best.basis(lc.tab)
> lc.bb
Wavelet Packet Transform of lc
Wavelet: s8
Length of series: 1024
Number of levels: 6
Boundary correction rule: zero
Crystals:
w5.0 w6.2 w6.3 w6.4 w6.5 w5.3 w6.8 w6.9 w6.10 w6.11
   ... (60 bases)
```

The best basis is composed of 61 different crystals. Plot the location
of these crystals in the table, resulting in figure 7.12 with:

```
> pgrid.plot(lc.bb)
```

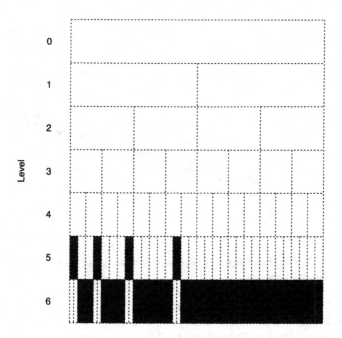

FIGURE 7.12. The minimum entropy basis for a linear chirp, selected using the
Coifman and Wickerhauser best basis algorithm.

Compute the time-frequency plot for the best basis with:

```
> time.freq.plot(lc.bb)
> abline(a=0, b=.25*512, lwd=2)
```

The result is shown in figure 7.13. The best basis does a better job
of localizing the frequency behavior of the chirp than the resolution
level 4 wavelet packet transform; compare with figure 7.10.

7.4.2 Computing the Best Level

The **best.level** function finds the best resolution for a wavelet
packet transform at a single level. The best level for the linear chirp
signal is level 6:

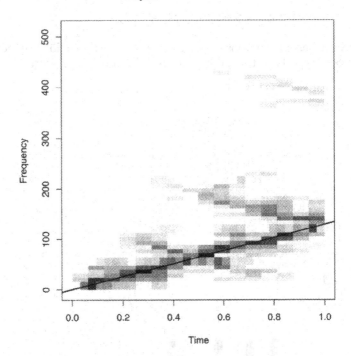

FIGURE 7.13. The time-frequency plot for the "best basis" wavelet packet transform of a linear chirp.

```
> lc.bl <- best.level(lc.tab)
> lc.bl
Wavelet Packet Transform of lc
Wavelet: s8
Length of series: 1024
Number of levels: 6
Boundary correction rule: zero
Crystals:
w6.0 w6.1 w6.2 w6.3 w6.4 w6.5 w6.6 w6.7 w6.8 w6.9
 ... (64 bases)
```

Computationally, searching for a best level is far easier than searching for a best basis, since you only need to compare J levels, where J is the maximum resolution level.

7.4.3 Using Tree Plots to Visualize Bases

A way to visualize the "relative entropy content" of the crystals is with the function `tree.plot`. Apply `tree.plot` to the best basis, the

level 4 basis, and the DWT for the linear chirp to produce figure 7.14
as follows:

```
> par(mfrow=c(1,3))
> tree.plot(lc.bb)
> tree.plot(lc.4)
> tree.plot(lc.dwt)
```

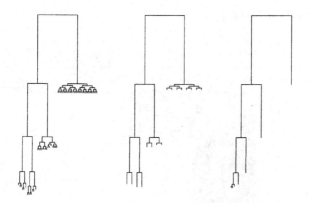

FIGURE 7.14. Tree plots for a linear chirp. Left: the best basis. Middle: level 4
basis. Right: DWT.

The length of the arcs between nodes indicates the entropy "sav-
ings" achieved by using the child nodes instead of the parent node.
Long arcs correspond to greater reduction in the entropy, and are
"important" splits. All of these tree plots are qualitatively similar.
This indicates that the additional entropy reduction achieved by the
best basis is not much greater than the DWT for a linear chirp.

Note: The exact definition of the length of the arc leading downwards from crystal $\mathbf{w}_{j,b}$ is given by

$$\max\left(0, \frac{1}{2}\left(E(\mathbf{w}_{j,b}) - \left(E^*_{j+1,2b} + E^*_{j+1,2b+1}\right)\right)\right) \qquad (7.8)$$

where the $E^*_{j,2b}$ are the minimum costs of the subtrees. They are
computed by the backward recursions

$$E^*_{j,2b} = \min\left(E(\mathbf{w}_{j,2b}), E^*_{j+1,2b} + E^*_{j+1,2b+1}\right)$$

where $E^*_{J,b} = E(\mathbf{w}_{J,b})$ at leaf nodes. For optimal bases, the arcs given
by (7.8) are guaranteed to be strictly positive.

7.4.4   Wavelet Packet Cost Tables

At the heart of the best basis algorithm is the "wavelet packet cost table," which is the table of costs $E(\mathbf{w}_{j,b})$. You can extract a cost table from a wavelet packet table using the `pcosts` function:

```
> lc.entropy <- pcosts(lc.tab)
```

Plot the costs to produce figure 7.15:

```
> plot(lc.entropy)
```

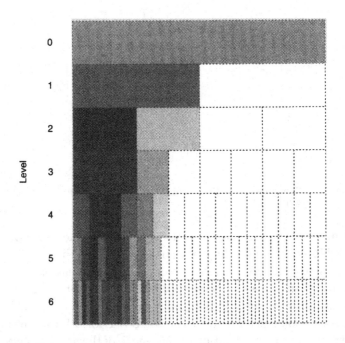

FIGURE 7.15. Plot of the wavelet packet cost table for a linear chirp.

The rectangles in figure 7.15 correspond to the normalized costs $2^j E(\mathbf{w}_{j,b})$. The factor 2^j corrects for the fact that different crystals have different numbers of coefficients. The "high cost" regions tend to correspond to high energy spots in the table; compare figure 7.15 to the plot of the wavelet packet table shown in figure 7.4.

A number of different cost functions are provided:

entropy: Define the cost measure

$$E_{j,b}^{\text{entropy}} = \sum_k \widetilde{w}_{j,b,k}^2 \log \widetilde{w}_{j,b,k}^2 \qquad (7.9)$$

where $\widetilde{w}_{j,b,k} = w_{j,b,k}/ \parallel \mathbf{w}_{0,0} \parallel_2$ (here $\parallel \cdot \parallel_2$ is the L_2 norm). Minimizing $E_{j,b}^{\text{entropy}}$ is the same as minimizing entropy. Entropy is commonly used in data compression and statistical estimation problems. This is the default cost function.

threshold: The threshold cost function counts the number of coefficients greater than a certain threshold:

$$E_{j,b}^{\text{thresh}} = \sum_k 1_{[|w_{j,b,k}|>t]}$$

where 1_A is the indicator function for the set A and t is a threshold. By default, the threshold is the median of the absolute value of all coefficients in the table. Minimizing threshold is useful for data compression schemes in which only the largest coefficients are retained.

sure: SURE is "Stein's Unbiased Risk Estimate" and the SURE cost function is defined by

$$E_{j,b}^{\text{SURE}} = \sigma^2 \left(n - 2 \sum_k 1_{[|w_{j,b,k}| \le t_j \sigma]} + \sum_k \min((w_{j,b,k}/\sigma)^2, t_j^2) \right).$$

SURE is useful for selecting optimal bases for extracting signals from noisy data [Don93b]. The parameter $t_n = \sqrt{2 \log_e(n \log_2 n)}$ is the shrinkage threshold and σ is a scale parameter for the noise (see chapter 6).

lp: The L_p norm cost function is defined by

$$E_{j,b}^{\text{lp}} = \left(\sum_k |w_{j,b,k}|^p \right)^{1/p}$$

where $p < 2$ (since all orthogonal transforms preserve energy, the costs for $p = 2$ are the same).

The pcosts function is used to compute a different cost function. For example, compute and plot the costs for the linear chirp signal using the threshold cost function as follows:

```
> lc.thresh <- pcosts(lc.tab, cost.fun="threshold")
> plot(lc.thresh)
```

The threshold costs are displayed in figure 7.16   You can supply

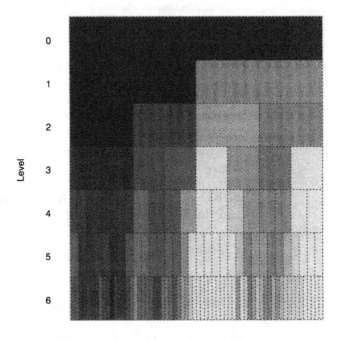

FIGURE 7.16. Plot of the wavelet packet cost table for a linear chirp using the threshold cost function.

a different cost function to the best basis function to get a different best basis:

```
> lc.bb1 <- best.basis(lc.tab, costs=lc.thresh)
> lc.bb1
Wavelet Packet Transform of lc
Wavelet: s8
Length of series: 1024
Number of levels: 6
```

```
Boundary correction rule: zero
Crystals:
w5.0 w5.1 w6.4 w6.5 w6.6 w6.7 w6.8 w6.9 w6.10 w6.11
... (52 bases)
```

Figure 7.17, which compares this basis to the "entropy" basis, is created as follows:

```
> par(mfrow=c(1,2))
> pgrid.plot(lc.bb)
> pgrid.plot(lc.bb1)
```

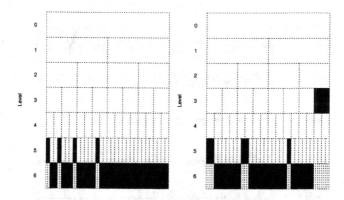

FIGURE 7.17. Comparison of best bases for different cost functions. Left: the entropy cost function. Right: the **threshold** cost function.

7.5   Wavelet Packet Analysis of a Speech Signal

In this section, you will use the wavelet packet functions to analyze a segment of a **speech.had** signal, which is the digitized signal of a speaker saying the word "had." Segment and plot the segmented signal as follows:

```
> speech <- speech.had[513:1024]
> speech <- speech - mean(speech)
> plot(speech, type="l")
```

The speech signal is shown in figure 7.18 Now compute and plot the wavelet packet table shown in figure 7.19:

```
> speech.tab <- wp.table(speech)
> plot(speech.tab)
```

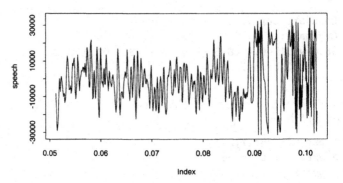

FIGURE 7.18. A segment of a speech signal.

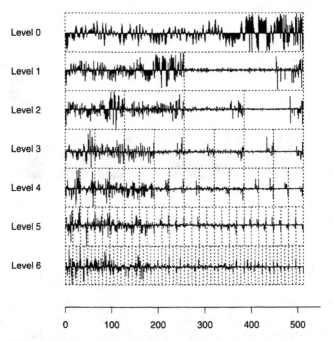

FIGURE 7.19. A wavelet packet table for the speech segment.

A good way to start an analysis is to apply the pre-packaged summary functions, `eda.plot` and `summary`. Apply `eda.plot` to the object `speech.tab` to produce figure 7.20:

```
> eda.plot(speech.tab)
```

The EDA plot for a wavelet packet tables shows:

1. A plot of the cost table.

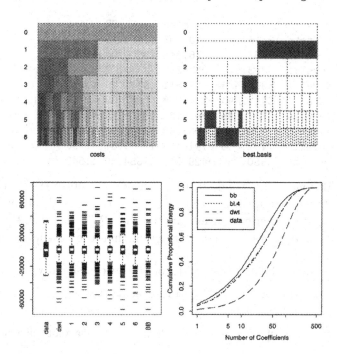

FIGURE 7.20. An EDA plot of a wavelet packet table for the speech segment.

2. A plot of the best basis.

3. Box plots of the coefficients for each resolution level, the DWT, and the best basis.

4. An energy plot for the best basis, best level (level 6 in this case), DWT, and the data (level 0).

The **summary** function produces a table of statistics comparing the different resolution levels, the DWT, and the best basis:

```
> summary(speech.tab)
                 Min        1Q    Median        3Q       Max
   data    -31163.40 -8891.402 -827.402  8069.598  34116.60
    dwt    -63130.98 -4173.943   -0.116  3934.043  61965.19
 level.1   -50666.73 -4669.828 -274.454  3934.043  44585.41
 level.2   -57601.31 -4596.596 -210.552  3668.892  53113.96
 level.3   -54284.91 -4286.966 -120.276  3384.584  74208.18
 level.4   -59080.00 -3898.699   -8.690  3826.279  66517.53
 level.5   -71892.76 -4981.497   25.285  4653.271  61965.19
 level.6   -63130.98 -5206.743 -296.623  4526.387  72451.58
     bb    -71892.76 -3868.120   32.727  3629.080  72451.58
```

	Mad	Mean	SD	Cost
data	12716.260	0.000	13561.21	5.502
dwt	5957.760	75.844	13561.00	4.798
level.1	6353.071	-257.384	13558.77	5.022
level.2	5996.536	-42.511	13561.15	4.943
level.3	5714.580	-301.348	13557.86	4.776
level.4	5727.685	-42.160	13561.15	4.776
level.5	7170.130	-163.948	13560.22	4.861
level.6	7162.546	-191.117	13559.86	4.796
bb	5550.159	-429.510	13554.40	4.543

In addition to the usual statistics, summary reports the cost of the basis (in this case, the entropy). For the speech signal, the best basis has the best energy concentration. The DWT has slightly higher entropy and poorer energy compaction than the best level and best basis.

Select and plot the best basis for the speech signal with:

```
> speech.bb <- best.basis(speech.tab)
> plot(speech.bb)
```

The result is shown in figure 7.21.   The wavelet packet coeffi-

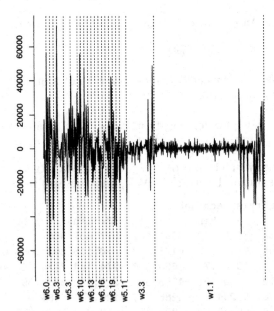

FIGURE 7.21. A plot of the best basis wavelet packet transform for the speech signal.

cients are plotted as vertical lines extending from zero. The coefficients are ordered roughly by frequency (i.e., oscillation and scale). In S+WAVELETS, this ordering is referred to as "depth.first." The dashed lines mark the different crystals (frequency bands). This plot is analogous to a periodiogram, although the coefficients are ordered by time within a crystal. You can see that the latter part of the signal has much more high frequency activity than the rest of the signal.

Produce an EDA plot for the best basis transform to obtain figure 7.22 as follows:

```
> eda.plot(speech.bb)
```

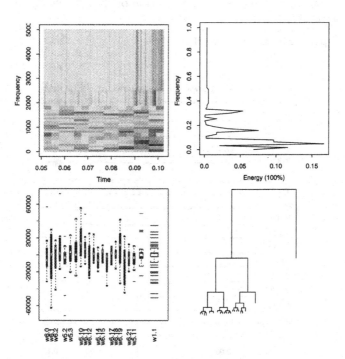

FIGURE 7.22. EDA plot of the best basis wavelet packet transform for the speech signal.

An EDA plot for a wavelet packet transform shows the following views:

1. A time-frequency plot.

2. The proportion of energy by frequency averaged over the entire signal.

3. Box plots of the coefficients by crystal. The crystals are ordered from low frequency to high frequency.

4. A tree plot.

The changing nature of the signal is reflected in the time-frequency plot; there is much more high-frequency activity in the latter portion of the signal. Two frequencies and their harmonics show up prominently, as indicated by the plot on the upper right. The tree plot shows that the best basis transform gains a moderate reduction in entropy by splitting the $w_{2,1}$ and $w_{3,1}$ and $w_{3,3}$ crystals to obtain finer frequency resolution than the DWT.

Plot the wavelet packet decomposition of the speech signal to produce figure 7.23 as follows:

```
> plot(decompose(speech.bb))
```

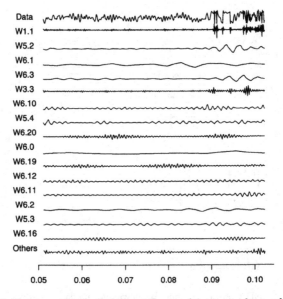

FIGURE 7.23. Decomposition of the speech signal into wavelet packet components.

By default, the signal components are ordered by energy, with the highest energy components plotted at the top. The highest energy component is $W_{1,1}$, probably because it represents the burst at the end of the signal. Besides the $W_{1,1}$ component, the best basis decomposition primarily consists of highly oscillatory components (unlike a DWT decomposition).

As with the DWT, you can analyze individual crystals in more depth. Compute the EDA plot for the $w_{1,1}$ crystal with:

```
> par(mfrow=c(2,2))
> eda.plot(speech.bb[["w1.1"]])
```

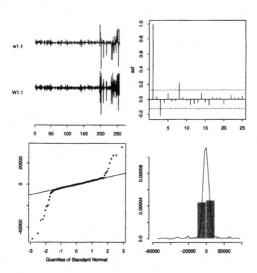

FIGURE 7.24. EDA plot of the $w_{1,1}$ crystal for the speech signal.

Figure 7.24 shows the result. The distribution of the coefficients is "long-tailed" and highly non-normal, resulting from the changing nature of the signal. The wavelet packet coefficients appear to display little autocorrelation. You should be careful, however, in assessing the autocorrelation in this case: the large coefficients at the end of the signal swamp any correlation present in the remainder of the signal.

8
Cosine Packet Analysis

This chapter discusses *cosine packet* analysis for one-dimensional signals. Cosine packet analysis is commonly known as *local cosine* analysis, and was first introduced by Ronald Coifman and Yves Meyer [CM91]. The term "cosine packets" was coined by David Donoho, another leading wavelets researcher, because cosine packet analysis is a mirror image of wavelet packet analysis. The difference is that localized cosine functions are used instead of wavelet packet functions.

In this chapter, you will learn about the following topics:

- Obtaining the discrete cosine transform (DCT) and block DCT with the functions `dct` and `block.dct` (see section 8.1).

- Creating and plotting cosine packet functions using the function `cosine.packet` and obtaining a block cosine packet transform using the function `block.cpt` (see section 8.2).

- Creating cosine packet tables using the `cp.table` function, selecting orthogonal cosine packet transforms (CPT) from a table, and automatic selection of transforms using the Coifman and Wickerhauser "best basis" algorithm (see section 8.4).

- Cosine packet analysis of a signal using visual data analysis tools (see section 8.5).

8.1   Discrete Cosine Transform

The discrete cosine transform (DCT) is an important and widely used tool in signal processing and image processing. The DCT is particularly valuable for coding and data compression applications, and it forms the core of the JPEG and MPEG algorithms [ISO91] for image compression. Like the discrete Fourier transform (DFT), the DCT defines an orthogonal transformation which maps a signal from the time domain to the frequency domain. However, the DCT is a *real-valued* transform, and does not involve complex numbers. The DFT is a complex-valued transform.

This section gives a brief introduction to the discrete cosine transforms implemented in S+WAVELETS. Algorithms for the DCT are discussed in section 13.2.1. For a comprehensive discussion of the DCT and applications involving the DCT, refer to the book by Rao and Yip, *Discrete Cosine Transform: Algorithms, Advantages, Applications* [RY90].

8.1.1   DCT Types

The *Fourier cosine transform* (FCT) of a signal $f(t)$ is given by

$$g(\omega) = \sqrt{\frac{2}{\pi}} \int_0^\infty f(t) \cos(\omega t) dt. \tag{8.1}$$

The discrete cosine transform is a discretized version of (8.1). There are four commonly used orthogonal discrete cosine transforms: DCT-I, DCT-II, DCT-III, and DCT-IV. In S+WAVELETS, you can compute two of these transforms, DCT-II and DCT-IV.

For a discrete signal $f_1, f_2, \ldots, f_n$, the DCT-II is defined as

$$g_k^{II} = \sqrt{\frac{2}{n}} s_k \sum_{i=0}^{n-1} f_{i+1} \cos\left(\frac{(2i+1)k\pi}{2n}\right) \quad k = 0, 1, \ldots, n-1. \tag{8.2}$$

The scaling factor s_k is defined by

$$s_k = \begin{cases} 1 & \text{if } k \neq 0 \text{ or } n \\ \frac{1}{\sqrt{2}} & \text{if } k = 0 \text{ or } n. \end{cases} \tag{8.3}$$

The inverse DCT-II is given by

$$f_{i+1} = \sqrt{\frac{2}{n}} \sum_{k=0}^{n-1} g_k^{II} s_k \cos\left(\frac{(2i+1)k\pi}{2n}\right) \quad i = 0, 1, \ldots, n-1. \tag{8.4}$$

The DCT-IV is defined by

$$g_k^{IV} = \sqrt{\frac{2}{n}} \sum_{i=0}^{n-1} f_{i+1} \cos\left(\frac{(2i+1)(2k+1)\pi}{4n}\right) \quad k = 0, 1, \ldots, n-1.$$
(8.5)

The inverse DCT-IV is given by

$$f_{i+1} = \sqrt{\frac{2}{n}} \sum_{k=0}^{n-1} g_k^{IV} \cos\left(\frac{(2i+1)(2k+1)\pi}{4n}\right) \quad i = 0, 1, \ldots, n-1.$$
(8.6)

Note: The DCT-IV does *not* include the constant term! Hence, all coefficients for the DCT-IV of a constant vector are non-zero. By contrast, only the first coefficient for the DCT-II of a constant vector is non-zero.

8.1.2 Computing the DCT

Use the **dct** function to compute the discrete cosine transform. As an example, we will look at the **ice** signal, which consists of underwater acoustic measurements of ice "cracking." Plot the **ice** signal, shown in figure 8.1, as follows:

```
> ts.plot(ice)
```

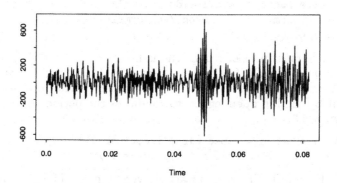

FIGURE 8.1. The **ice** signal of underwater acoustic noise.

By default, the function **dct** computes the DCT-II transform using the argument (**type=2**). Use the argument **type=4** to get the DCT-IV transform. Compute the DCT-II of the **ice** signal with:

```
> ice.dct <- dct(ice)
```

Now plot the absolute value of the coefficients of the DCT-II on a
decibel scale ($10 \log_{10}$) as follows:

```
> plot(10*log10(abs(ice.dct)[-1]),type="l")
```

This results in figure 8.2.   The first element of the DCT is propor-

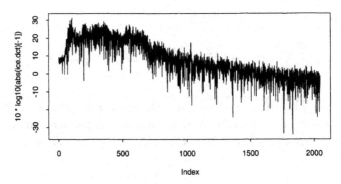

FIGURE 8.2. The DCT-II of `ice` signal plotted on a decibel scale.

tional to the absolute value of the mean of the signal, and it is not
plotted because it often is much bigger than the remaining elements
of the DCT.

You can invert the DCT using the argument `inv=T` of the `dct`
function:

```
> ice.recon <- dct(ice.dct, inv=T)
> vecnorm(ice-ice.recon)/vecnorm(ice)
[1] 4.280761e-15
```

You get almost perfect reconstruction.

Compare the DCT-II to the periodogram, which is essentially
given by the modulus of the discrete Fourier transform (DFT). The
S-PLUS function `spec.pgram` to computes the periodogram plotted
in figure 8.3:

```
> ice.pgram <- spec.pgram(ice,taper=0)
> plot(ice.pgram$spec[-1], type="l")
```

The DFT and DCT are qualitatively very similar. The periodogram,
however, has only half as many coefficients because the phase information is lost in taking the modulus of the transform.

8.1.3   Block DCT

The discrete cosine transform, like the discrete Fourier transform, is
not localized in time. This means that the DCT captures the average

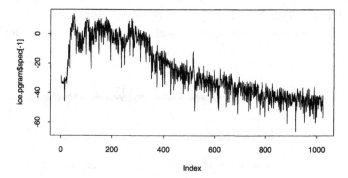

FIGURE 8.3. The periodogram of ice signal, which is the modulus of the discrete Fourier transform.

frequency behavior of a signal over the entire time period, and any information about local behavior (e.g., jumps) is lost. To capture local behavior of a signal, you can split the signal into blocks and analyze each block separately.

One way to split a signal is into dyadic blocks. For a dyadic level J, there are 2^J data blocks:

$$
\begin{aligned}
\text{Interval } 1 &= (f_1, f_2, \ldots, f_{n_J}) \\
\text{Interval } 2 &= (f_{1+n_J}, f_{2+n_J}, \ldots, f_{2n_J}) \\
\vdots &= \vdots \\
\text{Interval } 2^J &= (f_{n-n_J+1}, f_{n-n_J+2}, \ldots, f_n)
\end{aligned}
$$

where $n_J \equiv n/2^J$.

Use the function block.dct to compute the DCT over dyadic blocks of a signal. The argument n.level specifies the dyadic blocking factor J. Apply block.dct to compute the DCT for the ice signal split into $2^3 = 8$ blocks as follows:

```
> ice.block <- block.dct(ice, n.level=3)
```

Create a stack plot of the ice.block object shown in figure 8.4 with:

```
> stack.plot(ice.block, transform="abs")
```

The top plot shows the data (labeled "idct") with grid lines corresponding to the analysis blocks. The subsequent plots show the DCT coefficients. The argument transform="abs" causes the absolute value of the coefficients to be plotted. The block containing the "ice click" has markedly different frequency characteristics from the other blocks. Also, the latter part of the series appears to have somewhat different frequency characteristics from the beginning portion.

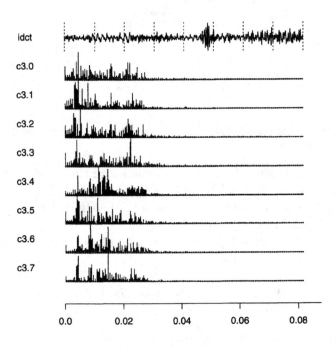

FIGURE 8.4. The DCT-II of ice signal split into 8 equal sized blocks. The top plot displays the original time series and the 8 blocks. The subsequent plots show the absolute value of the DCT-II coefficients for each of the blocks.

The block DCT has proven useful in many applications. One drawback of the block DCT is the abrupt "cutoff" implicit in dividing the signal into disjoint blocks. This can cause undesirable block effects, such as "Gibbs phenomena" (section 9.4 gives an example of block effects in the two-dimensional DCT). To avoid the problems caused by the abrupt cutoff, Coifman and Meyer [CM91] recently introduced a new type of localized cosine transform with smooth cutoffs (tapers). These transforms are discussed in the next section.

8.2   Cosine Packets

Cosine packet analysis, introduced by Ronald Coifman and Yves
Meyer [CM91], is a generalization of the discrete cosine transform
(DCT). Cosine packet analysis is commonly called *local cosine* analysis. Like the block DCT, a cosine packet transform (CPT) is based
on cosine functions which are *localized* in time. The key property
which distinguishes cosine packets from the block DCT is the use of
smooth basis functions.

A cosine packet function is obtained by damping a cosine function
down to zero on an interval I using a *taper function* or *bell function*
B_I. Type II and type IV cosine packet functions with frequency k
defined on the interval $I = [\alpha, \beta]$ are given by

$$C_k^{II}(t) = \sqrt{\frac{2}{\Delta_I}} B_I(t) \cos\left(\pi k(t - \alpha)/\Delta_I\right) \qquad (8.7)$$

$$C_k^{IV}(t) = \sqrt{\frac{2}{\Delta_I}} B_I(t) \cos\left(\pi (k + 1/2)(t - \alpha)/\Delta_I\right) \qquad (8.8)$$

where $\Delta_I = \beta - \alpha$.

The cosine packet obtained with the "boxcar" tapering function B_I
leads to the block DCT. The discontinuous cut-off at the boundaries
of the blocks can cause undesirable artifacts. The primary advantage
of smooth cosine packet functions is to avoid these artifacts.

In order to define orthogonal transforms, a very special kind of
tapering function is needed. The tapers extend *beyond* the ends of the
interval for which the cosine packet is defined. Create and plot a pair
of type II cosine packet functions defined on the interval $I = [0, 1]$
with frequency $k = 5$, one with the "boxcar" tapering function and
the other with the default (smooth) tapering function:

```
> par(mfrow=c(1,2))
> xlim <- c(-.5,1.5)
> plot(cosine.packet(frequency=5, taper="boxcar"),
+       xlim=xlim)
> plot(cosine.packet(frequency=5),xlim=xlim)
```

The result is plotted in figure 8.5. The tapered cosine packets (solid
line) are plotted with the tapering function (dashed line). The boxcar cosine packet is discontinuous while the default cosine packet
smoothly decays to zero. The smooth cosine packet spans $[-.5, 1.5]$,
extending beyond the original $[0, 1]$ interval.

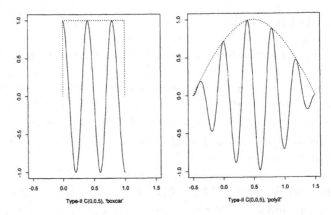

FIGURE 8.5. A cosine packet with frequency $k = 5$. Left: using a discontinuous "boxcar" taper function. Right: using a smooth taper function.

Cosine packet analysis parallels wavelet packet analysis—hence the name "cosine packet." Cosine packets and wavelet packets share many of the same data structures (e.g., packet tables) and functions (e.g., best.basis). This section discusses the functions which underlie the transform. The remaining sections present the basics of cosine packet analysis.

8.2.1   Special Tapering Functions

Very special tapering functions B_I are needed for the cosine packets defined by (8.7) and (8.8). Because the cosine packet functions overlap, only certain types of tapers preserve orthogonality in cosine packet analysis (see [AWW92, Wic94a]). S+WAVELETS offers 7 different tapers B_I which preserve orthogonality:

boxcar, poly1, poly2, poly3, poly4, poly5, trig

The boxcar taper is discontinuous, and is used for the block DCT. The poly1 is a polynomial taper with one continuous derivative. Likewise, the poly2, ..., poly5 are polynomial tapers with 2–5 continuous derivatives. The trig taper is based on trigonometric functions, and is the smoothest taper available in S+WAVELETS. The default taper in S+WAVELETS is the poly2 taper. For details on these tapers, turn to section 13.2.3.

The tapering function B_I defined for an interval $I = [\alpha, \beta]$ has the form

$$B_I(t) = \begin{cases} 0 & t \leq \alpha - \delta_\alpha \\ B\left((t - \alpha - \delta_\alpha)/(2\delta_\alpha)\right) & \alpha - \delta_\alpha < t < \alpha + \delta_\alpha \\ 1 & \alpha + \delta_\alpha \leq t \leq \beta - \delta_\beta \\ B\left((\beta + \delta_\beta - t)/(2\delta_\beta)\right) & \beta - \delta_\beta < t < \beta + \delta_\beta \\ 0 & t \geq \beta + \delta_\beta \end{cases} \qquad (8.9)$$

where B is one of the above tapering functions and $\Delta_I = \beta - \alpha$. The parameter $\delta_\alpha = \epsilon_\alpha \Delta_I$ defines the tapering region for the left hand side of the interval. Likewise, $\delta_\beta = \epsilon_\beta \Delta_I$ defines the tapering region for the right hand side.

Use the arguments dct.type, taper.type, and epsilon to select the DCT type, tapering type, and tapering regions. For example, the following expressions create the collection of cosine packets displayed in figure 8.6:

```
> par(mfrow=c(2,2))
> xlim <- c(-.5,1.5)
> plot(cosine.packet(freq=5,taper="poly1"))
> plot(cosine.packet(freq=5,taper="poly3"))
> plot(cosine.packet(freq=5,taper="trig",dct.type=4))
> plot(cosine.packet(freq=5,epsilon=c(.2,.5)),
+        xlim=xlim)
```

8.2.2 Dyadic Cosine Packet Families

A cosine packet can be defined on any interval $[\alpha, \beta]$. To construct orthogonal transforms (e.g., the block DCT), the cosine packets you will use to analyze your data in S+WAVELETS are restricted to dyadic blocks of your sampling interval $I = [i_0, i_1]$.

The j-bth block $I_{j,b}$ of the interval $I = [i_0, i_1]$ is defined by

$$I_{j,b} = [i_0 + b2^{-j}\Delta_I, i_0 + (b+1)2^{-j}\Delta_I]$$

where $\Delta_I = i_1 - i_0$.

A family of cosine packets $C_{j,b,k}$ has three parameters: $j = $ level, $b = $ block within level, and $k = $ frequency. For a sampling interval of $[0, 1]$, type II and type IV cosine packet families are defined by the

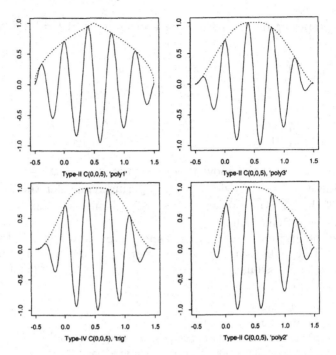

FIGURE 8.6. Different types of cosine packet functions: Top left: the `poly1` taper. Top right: the `poly3` taper. Bottom left: the `trig` taper. Bottom right: the `poly2` taper with $\delta_\alpha = .2$.

following two equations:

$$C^{II}_{j,b,k}(t) = 2^{(j+1)/2} B_{j,b}(t) \cos\left(2^j \pi k(t - b/2^j)\right) \qquad (8.10)$$

$$C^{IV}_{j,b,k}(t) = 2^{(j+1)/2} B_{j,b}(t) \cos\left(2^j \pi (k + 1/2)(t - b/2^j)\right) \quad (8.11)$$

We plot some members $C^{II}_{j,b,k}(t)$ of an orthogonal family of cosine packets in figure 8.7 with:

```
> par(mfrow=c(2,2))
> xlim <- c(-.2,1.2)
> plot(cosine.packet(level=3, block=0, freq=1),
+       taper=F, xlim=xlim)
> plot(cosine.packet(level=3, block=5, freq=5),
+       taper=F, xlim=xlim)
> plot(cosine.packet(level=3, block=3, freq=3),
+       taper=F, xlim=xlim)
> plot(cosine.packet(level=3, block=0, freq=7),
```

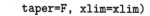

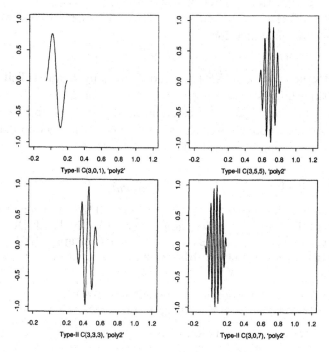

FIGURE 8.7. Members $C_{j,b,k}^{II}(t)$ of an orthogonal family of cosine packets.

8.3 Cosine Packet Transforms

In cosine packet analysis, a signal $f(t)$ is represented by a sum of orthogonal cosine packet functions $C_{j,b,k}(t)$ at different frequencies, blocks, and levels:

$$f(t) \approx \sum_{j} \sum_{b} \sum_{k} c_{j,b,k} C_{j,b,k}(t) \tag{8.12}$$

where $c_{j,b,k}$ are the cosine packet coefficients. The range of the summation for j and b is chosen so that the blocks are non-overlapping. This ensures that the cosine packet functions are orthogonal. The theoretical cosine packet coefficients are given by the integral

$$c_{j,b,k} = \int C_{j,b,k}(t) f(t) dt. \tag{8.13}$$

An orthogonal cosine packet transform (CPT) is defined as the mapping from the n signal values $\mathbf{f} = (f_1, f_2, \ldots, f_n)'$ to n cosine packet coefficients $\mathbf{c} = (c_1, c_2, \ldots, c_n)'$. The CPT is equivalent to the matrix multiplication

$$\mathbf{c} = \mathbf{Cf}.$$

As with the DWT and WPT, you don't actually do an explicit matrix operation, but use a fast algorithm; see section 13.2.

There are many possible cosine packet transforms. In this section, you will learn about one particular type of CPT. Section 8.4 shows how to select more general CPTs.

8.3.1 Block Cosine Packet Transform

The block cosine packet transform (block CPT) partitions the original sampling interval into dyadic blocks, like the block DCT. However, the block CPT can use one of the smooth tapers (poly1-poly5 or trig). Use the function block.cpt to compute a block CPT for a linear chirp signal with:

```
> lc <- make.signal("linchirp", n=1024)
> lc.cpt3 <- block.cpt(lc, n.level=3)
```

Then obtain figure 8.8 as follows:

```
> stack.plot(lc.cpt3)
```

The top plot shows the original signal with grid lines marking the 8 blocks. The transform coefficients for the 8 blocks are shown below. The increasing frequency of the chirp is shown very clearly by the coefficients.

Warning: Unlike in wavelet analysis, for most cosine packet analysis functions, the length n of the signal must be divisible by 2^J. If the length of your signal is not divisible by 2^J, you must either subset your signal or pad extra values to perform a cosine packet analysis.

Use the function nice.n to find the biggest integer not greater than n which is divisible by 2^J and the smallest integer not less than n which is divisible by 2^J. See page 128 of section 7.3.6 for an example of how to use the nice.n function.

8.3.2 Cosine Packet Transform Objects

Here is the cosine packet transform for the linear chirp:

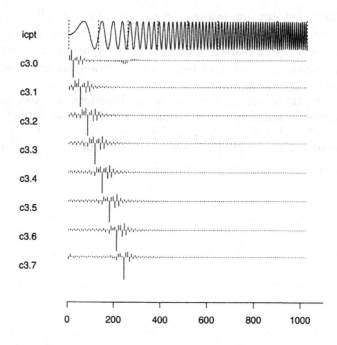

FIGURE 8.8. A block CPT of a linear chirp using the default smooth polynomial taper of degree 2.

```
> lc.cpt3
Block Cosine Packet Transform of lc
Number of Blocks: 8
Length of series:   1024
Number of levels: 3
Boundary extension rule: periodic
DCT Type: 2
Taper function type: poly2
Length of taper: 64
Crystals: c3.0 c3.1 c3.2 c3.3 c3.4 c3.5 c3.6 c3.7
```

By default, CPT uses the DCT-II type cosine packets, a polynomial taper of degree 2, and the **periodic** boundary extension rule. The default taper length is taken to be half the length of the block size, or $n/2^{J+1} = 1024/2^{3+1} = 64$. This means that the left taper is applied to the left 50% of the data in the the block and the right taper is applied to the right 50% of the data in the the block.

You can change the default parameters to a cosine packet analysis with the optional arguments **dct.type**, **taper**, **n.taper**, and boundary. Set dct.type=2 or 4, corresponding to DCT-II or DCT-IV. There are 7 different tapers to choose from; see section 8.2.1. The taper length n.taper can be set to any value less than or equal to the default. The different boundary extension rules are discussed in section 13.2.4.

The crystal names for a cosine packet transform are analogous to the names for a wavelet packet transform with the "w" replaced by a "c." Crystal c3.4 of the lc.cpt3 object corresponds to the transform of the fifth signal block. Plot this crystal to produce figure 8.9 with:

```
> plot(lc.cpt3[["c3.4"]], type="h")
```

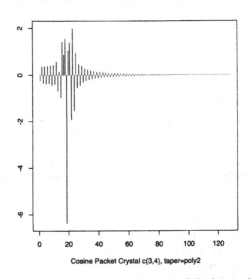

FIGURE 8.9. The c3.4 of the level 3 block CPT of the linear chirp signal.

8.3.3 Inverting and Decomposing CPTs

Virtually all of the functions which work for wavelet packet transforms also work for cosine packet transforms. The **reconstruct** function is used to invert a CPT:

```
> lc.recon <- reconstruct(lc.cpt3)
> vecnorm(lc.recon-lc)/vecnorm(lc)
[1] 4.757808e-16
```

The relative error of the reconstruction of the linear chirp is smaller than for wavelet packet transforms (see page 122).

Note: You can also compute the inverse transform using the icpt function.

As in wavelet analysis, you can decompose your signal into components based on the transform. For the level 3 block CPT transform, the decomposition is

$$f(t) = \mathbf{C}_{3,0}(t) + \mathbf{C}_{3,1}(t) + \cdots + \mathbf{C}_{3,7}(t)$$

where the signal components are given by

$$\mathbf{C}_{j,b}(t) = \sum_k c_{j,b,k} C_{j,b,k}(t).$$

Use the decompose function to compare the decomposition for the smooth CPT with that for the block DCT (which uses the boxcar taper):

```
> par(mfrow=c(1,2))
> plot(decompose(lc.cpt3, order="time"))
> plot(decompose(block.dct(lc,n.level=3), order="time"))
```

The resulting plot is shown in figure 8.10.   For the decomposition

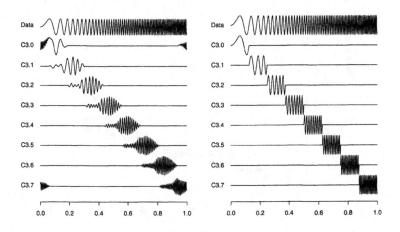

FIGURE 8.10. Decomposition of the linear chirp into cosine packet components. Left: using a smooth taper. Right: using a boxcar taper.

with the smooth taper, the signal components overlap. This is because the tapers extend beyond the ends of the blocks—in this case, by 64 sample values, or half the block size. The signal components for the boxcar taper are disjoint. The argument `order="time"` orders the blocks in the stack plot by time; by default, the blocks are ordered by "energy".

8.3.4 Time-Frequency Plots with CPTs

A cosine packet transform leads directly to a time-frequency plot like that produced for the wavelet packet transforms (see section 7.3.6). The function `time.freq.plot` computes the time-frequency plane plot for the linear chirp signal using the level 3 cosine packet transform as follows:

```
> time.freq.plot(lc.cpt3)
> abline(a=0, b=.25*512, lwd=2)
```

The resulting plot is shown in figure 8.11. The level 3 CPT does a

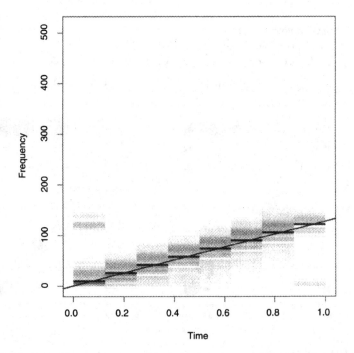

FIGURE 8.11. Time-frequency plane plot of the linear chirp using a level 3 cosine packet transform.

good job of isolating the chirp, identifying the linear chirp more clearly than any of the wavelet packet transforms. Compare figure 8.11 to the wavelet packet transform time-frequency plane plots in figure 7.10 (level 4 WPT), figure 7.11 (DWT), and figure 7.13 (best basis).

By default, the color or gray level of each box corresponds to the square root of the absolute value of a cosine packet coefficient $c_{j,b,k}$. The horizontal and vertical center of a box is located roughly at the time and frequency *center* of the associated cosine packet function. The width and height of the box is given by the time and frequency *bandwidth*. For a signal sampled at times $t_0, t_0+\Delta_t, \ldots, t_0+(n-1)\Delta_t$, the center of the box for coefficient $c_{j,b,k}$ is

$$(x,y) = \left(t_0 + n\Delta_t(b+.5)/2^j\right), (k+.5)2^j/n).$$

The width and height of the boxes are given by

$$(\Delta x, \Delta y) = (n\Delta_t/2^j, 2^j/n).$$

8.4 Cosine Packet Tables

In section 8.3.1, you learned how to use the `block.cpt` function to obtain a cosine packet transform with equally sized analysis blocks. For signals which change over time, this may not be the best way to choose the blocks. Instead, you might want to choose short blocks to analyze parts of the signal with rapidly changing features and longer blocks to analyze stationary parts of a signal. A rich family of cosine packet transforms (CPTs) can be extracted from a *cosine packet table*.

Suppose you have n sampled signal values $\mathbf{f} = (f_1, f_2, \ldots, f_n)'$, where n is a multiple of 2^J. The cosine packet table has $J+1$ "blocking" levels, where J is the finest level of blocking. At each level j, the signal is blocked into 2^j dyadic intervals, each of length $n/2^j$. The $(J+1) \times n$ table of coefficients is obtained by stacking the $J+1$ levels on top of one another.

Table 8.1 shows the layout of a cosine packet table with 3 resolution levels. The level 0 coefficients $\mathbf{c}_{0,0} = (c_{0,0,1}, c_{0,0,2}, \ldots, c_{0,0,n})$ in the table are equal to the level 0 block CPT. The level 1 crystals $\mathbf{c}_{1,0}$ and $\mathbf{c}_{1,1}$ correspond to the level 1 block CPT. In general, the level j coefficients correspond to the level j block CPT, and there are 2^j blocks $b = 0, 1, \ldots, 2^j - 1$. Each block has $n/2^j$ frequencies $k = 0, 1, \ldots, n/2^j - 1$.

Level 0	$c_{0,0}$							
Level 1	$c_{1,0}$				$c_{1,1}$			
Level 2	$c_{2,0}$		$c_{2,1}$		$c_{2,2}$		$c_{2,3}$	
Level 3	$c_{3,0}$	$c_{3,1}$	$c_{3,2}$	$c_{3,3}$	$c_{3,4}$	$c_{3,5}$	$c_{3,6}$	$c_{3,7}$

TABLE 8.1. Cosine packet table with 3 resolution levels.

A cosine packet table mirrors a wavelet packet table with frequency and time reversed. In a cosine packet table, the blocks within a level are ordered by time and the coefficients within a block are ordered by frequency. In a wavelet packet table, the blocks within a level are ordered by frequency and the coefficients within a block are ordered by time.

8.4.1 Computing a Cosine Packet Table

In S+WAVELETS, you can create a cosine packet table with the function **cp.table**. Create and plot a cosine packet table for a linear chirp as follows:

```
> lc <- make.signal("linchirp", n=1024)
> lc.cptab <- cp.table(lc)
> plot(lc.cptab)
```

The plot is shown in figure 8.12. Each level has $n = 1024$ coefficients, and is divided into blocks of length 2^j, indicated by dashed grid lines (see table 8.1). The cosine packet coefficient $c_{j,b,k}$ is plotted as a vertical line extending from zero. The coefficients in a given level are all plotted on the same vertical scale.

Here is a look at the **lc.cptab** object:

```
> lc.cptab
Cosine Packet Table for  lc
Length of series:  1024
Number of levels: 6
Boundary extension rule: periodic
DCT Type: 2
Taper function type: poly2
Length of taper: 8
```

A cosine packet table has the same defaults as a cosine packet transform; compare **lc.cptab** with the **lc.cpt3** object (see page 155). For a table with $J = 6$ levels, the length of the taper is given by $n/2^{J+1} = 1024/2^7 = 8$.

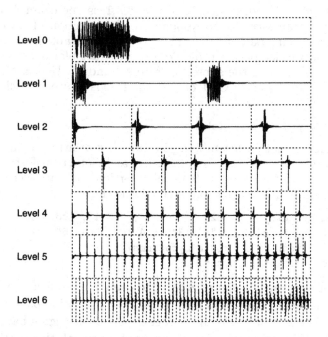

FIGURE 8.12. Cosine packet table for the linear chirp signal.

Note: The taper length is chosen to be the largest possible length for the shortest block in the packet table. Hence, the taper length is determined by the number of levels in the table. If you compute a cosine packet table with just 4 levels for the linear chirp signal, the taper length will be $1024/2^5 = 32$.

8.4.2 Selecting Cosine Packet Transforms from Tables

Orthogonal cosine packet approximations of the form (8.12) involve particular subsets of n coefficients from a cosine packet table. The rules for selecting general cosine packet transforms (CPTs) are the same as for choosing a wavelet packet transform from a wavelet packet table (see page 121):

1. Every column in the cosine packet table is covered by one crystal. This ensures the transform can be inverted to reconstruct the signal.

2. No column in the cosine packet table has more than one crystal. This ensures the transform is orthogonal.

The block CPT, discussed in section 8.3.1, is one example of an orthogonal transform. It corresponds to the selection of an entire row from a cosine packet table. The cosine packet table is organized just like a wavelet packet table. You can extract blocks or crystals from the table using the subset operators [and [[. Hence, another way to obtain the block CPT of a linear chirp is to use the following expression:

```
> lc.cpt3 <- lc.cptab[level=3]
```

As with wavelet packet tables, you can select arbitrary transforms using the subset operators. Here we select and plot a transform which uses wide blocks for the low frequency portion and narrower blocks for the high frequency portion of the linear chirp:

```
> mycpt.nms <- c("c2.0", "c2.1", "c3.4", "c3.5",
+          "c4.12", "c4.13", "c4.14", "c4.15")
> lc.mycpt <- lc.cptab[mycpt.nms]
> stack.plot(lc.mycpt)
```

The plot is shown in figure 8.13. The level 2 blocks in the lower frequency part of the signal have $1024/2^2 = 256$ cosine packet coefficients. The level 4 blocks in the higher part of the signal have only $1024/2^4 = 64$ coefficients. For the linear chirp signal, this particular CPT is analogous to the DWT, which has good frequency resolution but poor time resolution at low frequencies and poor frequency resolution but good time resolution at high frequencies.

Note: You can directly compute a cosine packet transform from a signal with the function cpt.

8.4.3 Best Basis Selection for Cosine Packets

The Coifman and Wickerhauser [CW92] "best basis" algorithm for selecting optimal bases (i.e., transforms) from wavelet packet tables applies in the same manner to cosine packet tables. You can use the functions best.basis and best.level to automatically select a transform from a cosine packet table. To visualize the transform choice, you can use the functions pgrid.plot and tree.plot. Section 7.4 gives background on best basis selection in S+WAVELETS.

The cosine packet best basis is selected based on a "cosine packet cost table," which is analogous to the wavelet packet cost table. The cost table is computed using the pcosts function. The cost functions, such as entropy, for wavelet packets can also be used for cosine packets. Turn to section 7.4.4 for a description of the cost

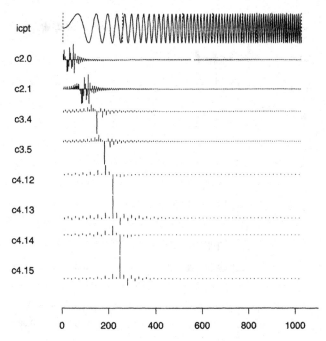

FIGURE 8.13. Cosine packet transform of a linear chirp using wide blocks for the low frequency portion and narrow blocks for the high frequency portion.

functions.

In this section, the ice signal is used to illustrate best basis cosine packet analysis. We perform a best basis analysis for the ice signal, which we previously analyzed using the block DCT in section 8.1.3. First, compute a cosine packet table for the de-meaned signal with:

```
> ice.tab <- cp.table(ice-mean(ice))
```

It is important to remove the mean to avoid possible problems with the best basis algorithm (the mean term in the DCT can dominate the "entropy" cost function). Now compute and plot a best basis to produce figure 8.14 as follows:

```
> ice.bb <- best.basis(ice.tab)
> stack.plot(ice.bb)
```

The best basis partition of the original signal is fine around the ice click, which involves a short burst of rapidly changing behavior. Other parts of the signal have relatively constant frequency characteristics, which is reflected by the best basis partition.

As with wavelet packets, you can access the packet cost (entropy)

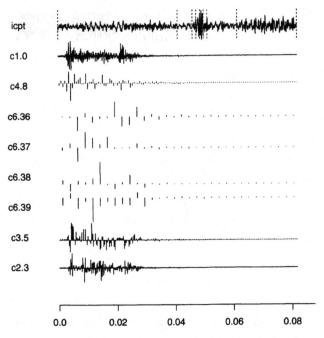

FIGURE 8.14. Best cosine packet basis for the ice signal.

table using the **pcosts** function and visualize the relative costs in a basis using the **tree.plot** function. Apply these functions to **ice.bb** to obtain figure 8.15 as follows:

```
> par(mfrow=c(1,2))
> plot(pcosts(ice.tab))
> tree.plot(ice.bb)
```

The ice click shows up very strongly in the finest scale—this corresponds with the partition selected by the best basis algorithm. There are significant "gains" by using finer partitions for modeling the ice click.

Compare the time-frequency plane plot based on the best basis with the plot based on the level 3 block DCT computed in section 8.1.3:

```
> par(mfrow=c(1,2))
> time.freq.plot(ice.bb)
> ice.block <- block.dct(ice, n.level=3)
> time.freq.plot(ice.block)
```

The plot is shown in figure 8.16. The best basis does a better job

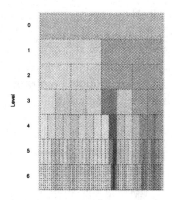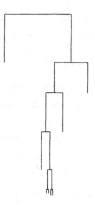

FIGURE 8.15. Entropy for the cosine packet table of the ice signal and tree plot of the best cosine packet basis for the ice signal.

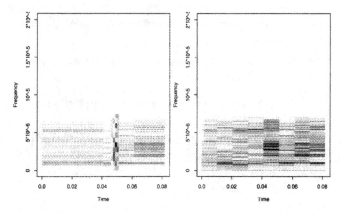

FIGURE 8.16. Time-frequency plane plot for the ice signal using the "best" cosine packet transform and a level 3 block DCT.

than the block DCT at localizing in time the ice click and localizing in frequency the ambient background noise.

8.5 Cosine Packet Analysis of a Speech Signal

To explore more about cosine packet analysis, and to compare and contrast it with wavelet packet analysis, we re-analyze the speech signal examined in section 7.5. The original speech segment was plotted in figure 7.18 on page 136. Compute and plot the cosine packet table for this signal as follows:

```
> speech <- speech.had[513:1024]
> speech <- speech - mean(speech)
> speech.cptab <- cp.table(speech-mean(speech))
> plot(speech.cptab)
```

As usual, we have de-meaned the signal before doing a cosine packet analysis. The cosine packet table is shown in figure 8.17.

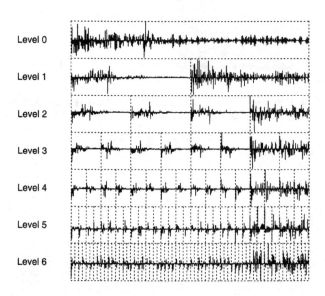

FIGURE 8.17. A cosine packet table for the speech segment.

Next we apply **eda.plot** to **speech.tab** to produce figure 8.18:

```
> eda.plot(speech.cptab)
```

An EDA plot for a cosine packet table shows:

1. A plot of the cost table.

2. A plot of the best basis.

3. Box plots of the data, the coefficients at each level, and the best basis.

4. The energy plot comparing the best basis to the best level, to the level 0, and to the data.

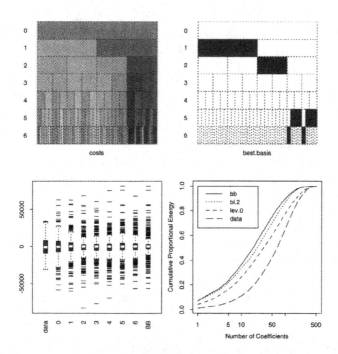

FIGURE 8.18. An EDA plot of a cosine packet table for the speech segment.

Now select and plot the best basis to produce figure 8.19 with:

```
> speech.cpbb <- best.basis(speech.cptab)
> stack.plot(speech.cpbb)
```

The best basis partition selects long segments for the beginning of the signal, which is relatively homogeneous, and short segments for the end of the signal, which contains apparently less homogeneous bursts.

Produce the EDA plot in figure 8.20 as follows:

```
> par(mfrow=c(2,2))
> eda.plot(speech.cpbb)
```

The EDA plot for CPTs shows a time-frequency plot (top left), the proportion of the energy of the signal by frequency plotted on a log scale (top right), box plots of the coefficients by crystal (bottom left), and a tree plot (bottom right).

The nature of the splits towards the end of the signal, which consists of both level 5 and level 6, may or may not correspond to actual features in the signal. In the best basis algorithm, there is no penalty

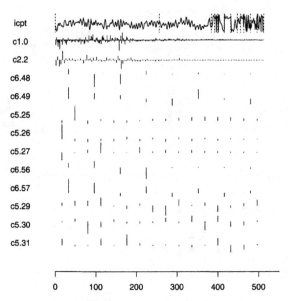

FIGURE 8.19. A stack plot of the best cosine packet basis for the speech signal.

for splitting. In an entirely random series, you are likely to see "random splitting." As shown by the tree plot in figure 8.20, these splits decrease the entropy of the transformed signal, but only by a small amount.

As with wavelet packets, you can access and analyze individual crystals. Extract and apply the `eda.plot` function to the first crystal $c_{1,0}$ to product figure 8.21 with:

```
> eda.plot(speech.cpbb[["c1.0"]])
```

This EDA plots shows:

1. A plot of the crystal and the reconstructed signal based on the crystal (top left).

2. The autocorrelation function (top right). At first glance, the autocorrelation function (ACF) would appear to indicate that the coefficients are not significantly correlated. However, the ACF is dominated by a few large coefficients, which mask the true correlation. To see this, use the `cor` function with the `trim=.1` to compute a robust estimate of lag one correlation.

3. The quantile–quantile plot of the empirical distribution versus a quantiles of a standard normal distribution (bottom left). The

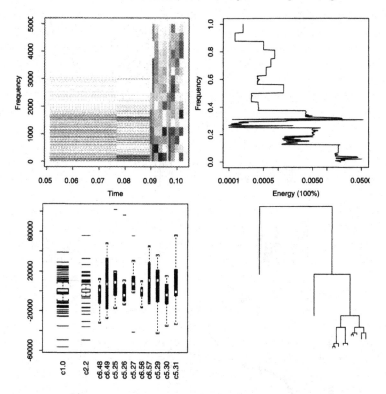

FIGURE 8.20. An EDA plot of the best cosine packet basis for the speech signal.

distribution is very long-tailed and highly non-normal.

4. The histogram and density estimate (bottom right) confirm that the distribution is long-tailed.

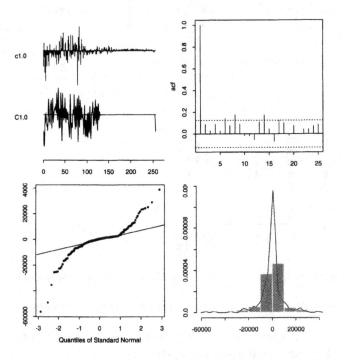

FIGURE 8.21. An EDA plot of the cosine packet crystal $c_{1,0}$ extracted from the best basis transform of speech signal. Top left: a plot of the crystal and the reconstructed signal based on the crystal. Top right: the autocorrelation function. Bottom left: the quantile-quantile plot of the empirical distribution versus a quantiles of a standard normal distribution. Bottom right: a histogram and density estimate of the distribution.

9

Wavelet Packet and Cosine Packet Analysis of Images

In this chapter, you will learn about wavelet packet and cosine packet analysis for images and matrices. You will learn how to do the following tasks:

- Create two-dimensional (2-D) wavelet packet and cosine packet objects with `wavelet.packet.2d` and `cosine.packet.2d` (section 9.1).

- Obtain 2-D wavelet packet transforms and 2-D cosine packet transforms with the functions `wpt.2d` and `cpt.2d` (section 9.2).

- Compute 2-D wavelet packet and cosine packet cost tables using the functions `wp.costs.2d` and `cp.costs.2d` and use "best basis" selection 2-D cost tables (section 9.3).

- Analyze a digital fingerprint image using 2-D wavelet packet and cosine packet functions (section 9.4).

9.1   2-D Wavelet and Cosine Packet Functions

You can extend wavelet packets and cosine packets to two-dimensions in exactly the same manner as for wavelets. Construct a 2-D wavelet packet by taking the tensor product of a *horizontal* 1-D wavelet packet function $W_{b_h}(x)$ and a *vertical* 1-D wavelet packet function $W_{b_v}(x)$:

$$W_{b_h, b_v}(x, y) = W_{b_h}(x) W_{b_v}(y). \tag{9.1}$$

The subscripts b_h and b_v correspond to the horizontal and vertical oscillation of the wavelet packet function. The 2-D wavelet packet $W_{4,2}(x, y)$ shown in figure 9.1 is created as follows:

```
> wp2d <- wavelet.packet.2d(wavelet="bs3.1",
+                  oscillation=c(4,2))
> plot(wp2d,eye=c(8,-4,24))
```

The argument `oscillation=c(4,2)` specifies the horizontal oscillation as 4 and the vertical oscillation as 2.

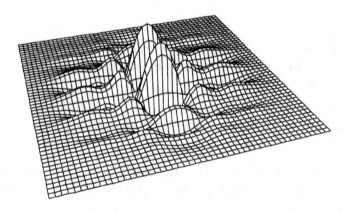

FIGURE 9.1. The 2-D wavelet packet function $W_{4,2}(x, y)$.

To get the full set of building block functions for 2-D wavelet packet analysis, you form scale and translation families of 2-D wavelets analogous to the 1-D case (see equation (7.1)). The general 2-D wavelet packet function is:

$$W_{(j_h, b_h, k_h),(j_v, b_v, k_v)}(x, y) = W_{j_h, b_h, k_h}(x) W_{j_v, b_v, k_v}(y)$$

where $W_{j_h,b_h,k_h}(x)$ and $W_{j_v,b_v,k_v}(y)$ are defined in (7.1). In addition to horizontal and vertical oscillation b_h and b_v, 2-D wavelet packets are parameterized by horizontal and vertical levels j_h and j_v and horizontal and vertical shifts k_h and k_v. Print the 2-D wavelet packet wp2d as follows:

```
> wp2d
2D Wavelet Packet Function:

Horizontal Function: Wavelet Packet Function:
Wavelet name: bs3.1
Oscillation: 4
Level: 0
Shift: 0

Vertical Function: Wavelet Packet Function:
Wavelet name: bs3.1
Oscillation: 2
Level: 0
Shift: 0
```

The level and shift parameters are changed with the arguments level and shift.

A 2-D cosine packet is also formed by a tensor product:

$$C_{k_h,k_v}(x,y) = C_{k_h}(x)C_{k_v}(y). \tag{9.2}$$

where $C_{k_h}(x)$ and $C_{k_v}(y)$ are 1-D cosine packet functions defined in (8.7) and (8.8). The cosine packet $C_{5,3}(x,y)$ has horizontal frequency $k_h = 5$ and vertical frequency $k_v = 3$. Create and plot the 2-D cosine packet $C_{5,3}(x,y)$ shown in figure 9.2 as follows:

```
> cp2d <- cosine.packet.2d(freq=c(5,3))
> plot(cp2d, J=5)
```

Analogous to 1-D cosine packets, the family of 2-D cosine packets is generated using dyadic blocks of your original image or matrix.

9.2   2-D Packet Transforms

To illustrate 2-D wavelet packet and cosine packet transforms, let us look at the **brain** image, which is a magnetic resonance (MR) image of Eve Riskin's brain. Eve Riskin is a professor at the University of Washington, and a leading researcher in data compression. We plot the image, which is shown in figure 9.3:

```
> image(brain)
```

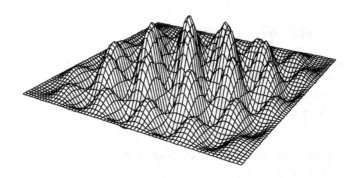

FIGURE 9.2. The 2-D cosine packet function $C_{5,3}(x, y)$.

9.2.1 Computing 2-D Wavelet Packet Transforms

For 1-D data, we have seen the wavelet packet transform using all wavelet packets at a given level in section 7.3.2. In this transform, all of the crystals (or subbands) have the same number of coefficients. To compute the analogous transform for 2-D data, use the function wpt.2d.

Compare the 2-D DWT with a level 3 wavelet packet transform for the brain image as follows:

```
> par(mfrow=c(1,2))
> brain.dwt <- dwt.2d(brain, n.level=3)
> brain.wpt <- wpt.2d(brain, n.level=3)
> plot(brain.dwt)
> plot(brain.wpt)
```

The result is shown in figure 9.4. The argument n.level=3 specifies the level in the transform.

The level 3 WPT splits the finer scale coefficients of the DWT into smaller blocks. For example, the d1-s1 block in the DWT has dimensions 128×128, and is broken up into 16 blocks in the WPT. These

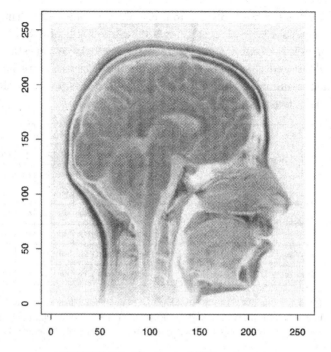

FIGURE 9.3. The **brain** MR image.

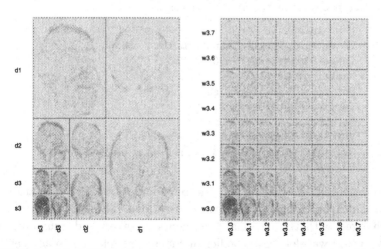

FIGURE 9.4. 2-D DWT (left) and a 2-D level 3 wavelet packet transform (right) of the **brain** image

blocks are w3.4-w3.0, w3.5-w3.0, ..., w3.7-w3.3, each of which has dimension 32×32. The coarse scale level 3 coefficients of the

DWT, s3-s3, s3-d3, d3-s3, and d3-d3, are identical to their WPT counterparts w3.0-w3.0, w3.1-w3.0, w3.0-w3.1, and w3.1-w3.1.

Figure 9.5 shows the complete mapping between crystals for the 2-D DWT and the level 3 WPT. The WPT crystals are labeled by oscillation number: the block labeled **0-3** corresponds to the WPT coefficient matrix w3.0-w3.3.

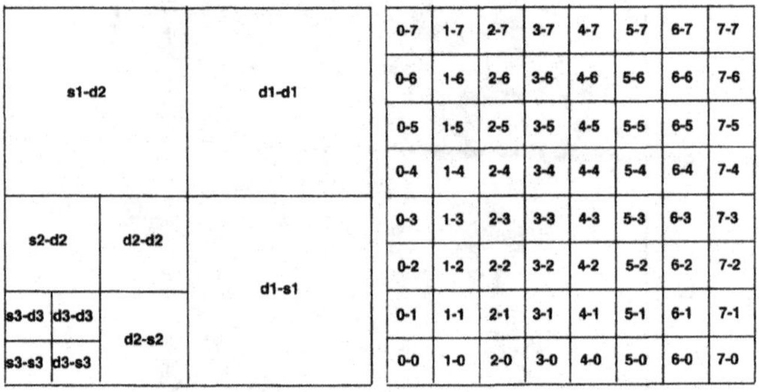
s1-d2
d1-d1
s2-d2
d2-d2
d1-s1
s3-d3
d3-d3
d2-s2
s3-s3
d3-s3
0-7 1-7 2-7 3-7 4-7 5-7 6-7 7-7
0-6 1-6 2-6 3-6 4-6 5-6 6-6 7-6
0-5 1-5 2-5 3-5 4-5 5-5 6-5 7-5
0-4 1-4 2-4 3-4 4-4 5-4 6-4 7-4
0-3 1-3 2-3 3-3 4-3 5-3 6-3 7-3
0-2 1-2 2-2 3-2 4-2 5-2 6-2 7-2
0-1 1-1 2-1 3-1 4-1 5-1 6-1 7-1
0-0 1-0 2-0 3-0 4-0 5-0 6-0 7-0

FIGURE 9.5. The 2-D DWT wavelet coefficient matrices (left) and the level 3 WPT coefficient matrices (right). The WPT crystals are labeled by oscillation number: the block marked **0-3** corresponds to the WPT coefficient matrix w3.0-w3.3.

Here is a look at the WPT object:

```
> brain.wpt
2D Wavelet Packet Transform for: brain
Image Dimensions:  256 by 256
Number of Levels:  3
Horizontal Wavelet:  s8
Vertical Wavelet:  s8
Horizontal Boundary Rule:  periodic
Vertical Boundary Rule:  periodic
Crystal names: w3.0-w3.0 w3.1-w3.0 w3.2-w3.0 ... (
64 crystals)
```

The default wavelet s8 and the periodic boundary rule are used for both horizontal and vertical directions. These defaults can be changed with the optional arguments wavelet and boundary. As with 2-D wavelets, the coefficient blocks are referred to as *crystals*. There are $2^3 \times 2^3 = 64$ crystals in a level 3 wavelet packet transform.

To extract a crystal from a WPT, the usual subscripting conventions apply. For example, we would compute an EDA plot for the w3.0-w3.3 crystal as follows:

```
> eda.plot(brain.wpt[["w3.0-w3.3"]])
```

The resulting plot shows the 32×32 block of coefficients (top left), the reconstructed image from these coefficients (top right), the 2-D autocorrelation function (ACF) (bottom left), and a histogram and density estimate of the coefficients (bottom right). Like its DWT counterpart s2-d2, the w3.0-w3.3 crystal represents horizontal edges in the original image. It also does a very good job of energy compaction, as indicated by the density plot.

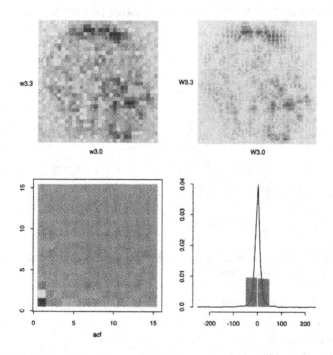

FIGURE 9.6. EDA plot for the wavelet packet crystal w3.0-w3.3 of **brain** image.

9.2.2 Computing 2-D Cosine Packet Transforms

The 2-D cosine packet transform is obtained by partitioning the image into dyadic blocks and transforming each block using a 2-D cosine transform. The same tapers can be used as in the 1-D setting:

boxcar, poly1, poly2, poly3, poly4, poly5, trig.

Compute and plot a 2-D level 3 block cosine packet transform of the **brain** image with:

```
> par(mfrow=c(1,2))
> brain.cpt <- cpt.2d(brain, n.level=3,
+      boundary="periodic")
> image(brain)
> pgrid.plot(crystal.names(brain.cpt),add=T,
+      dim=dim(brain))
> plot(brain.cpt)
```

The plot is shown in figure 9.7. The function pgrid.plot plots a grid corresponding to the blocking in the brain.cpt object. The argument add=T adds the grid to the existing plot of the brain image.

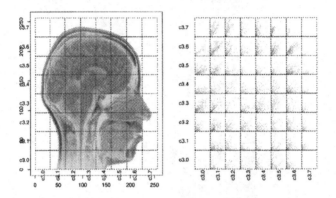

FIGURE 9.7. Left: the brain image divided into dyadic blocks. Right: the corresponding cosine packet transform.

Here is the brain.cpt object:

```
> brain.cpt
2D Local Cosine Transform for  brain
Image Dimensions:  256 by 256
Number of Levels:  3
Horizontal Taper Function:  poly2
Vertical Taper Function:  poly2
Horizontal Boundary Rule:  periodic
Vertical Boundary Rule:  periodic
Horizontal DCT-type:  2
Vertical DCT-type:  2
Crystal names: c3.0-c3.0 c3.1-c3.0 c3.2-c3.0 ... (
64 crystals)
```

The DCT-II transform, the periodic boundary extension rule, and a polynomial taper of degree 2 are used for both the horizontal and vertical directions. These can be changed with the optional arguments `dct.type`, `boundary`, and `taper`. For a discussion of these arguments, see chapter 8, Cosine Packet Analysis.

Selecting the `c3.0-c3.3` crystal and creating an EDA plot

```
> eda.plot(brain.cpt[["c3.0-c3.3"]])
```

lead to figure 9.8. A cosine packet crystal represents a spatial block

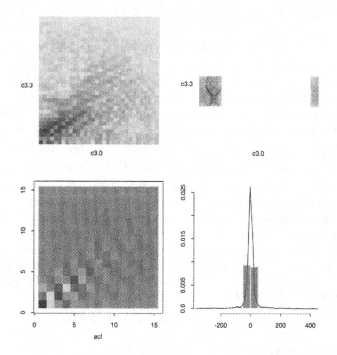

FIGURE 9.8. EDA Plot for the cosine packet crystal `c3.0-c3.3` of the `brain` image.

of the image. The largest coefficients extend diagonally in the cosine packet transform domain. This is because the original block contains a diagonal edge; see the reconstructed image in the top right of figure 9.8. In general, if the original image block has an edge, then the cosine packet coefficients also form an edge perpendicular to the original image.

Because a smooth taper is used, the reconstructed image extends beyond the boundaries of the block and even wraps around the image! It also does a very good job of energy compaction, as indicated by the density plot.

Warning: As in 1-D cosine packet analysis functions, the dimensions m and n of the image must both be divisible by 2^J. If not, you must either subset the image or pad extra values to perform a cosine packet analysis. Use the function `nice.n` to find the biggest integers not greater than m, n which are divisible by 2^J.

9.3   Best Basis Selection in 2-D

As in the 1-D case, there are many possible 2-D wavelet packet and cosine packet transforms. For wavelet packets, you can split a co-efficient block into four separate blocks, or merge four blocks into a single block. Similarly, for cosine packets, you can split or merge blocks to analyze smaller or larger blocks. Fortunately, the Coifman and Wickerhauser best basis algorithm extends to two-dimensions.

In the 1-D case, it is possible to compute a whole table of wavelet packet or cosine packet coefficients. In 2-D, this is no longer desirable because a table of coefficients often takes up too much storage; a table of J levels contains $J \times m \times n$ double precision numbers. For even a moderate image size such as 256×256, this can require several megabytes of storage.

Instead of storing the wavelet packet table coefficients, you can compute the "costs" using the function `wp.costs.2d`. Using the costs, you can find a best basis, just as in the 1-D case. Compute the wavelet packet costs for the **brain** image as follows:

```
> brain.wpc <- wp.costs.2d(brain, n.level=6)
> brain.wpc
Wavelet packet cost vector for brain
Cost function: entropy
Image Dimensions:  256 by 256
Number of Levels:  6
Horizontal Wavelet:  s8
Vertical Wavelet:  s8
Horizontal Boundary Rule:  periodic
Vertical Boundary Rule:  periodic
```

This computes a cost table of 6 levels using the default **s8** wavelet and `periodic` boundary conditions. The "entropy" cost function is used; other cost functions available for 2-D wavelet packet analysis include "threshold" and "L_p". The "SURE" cost function is not available. See section 7.4.4 for a discussion of these cost functions.

You can plot the costs one level at a time. Plotting the costs for levels 1–4 produces figure 9.9:

```
> par(mfrow=c(2,2))
> for(i in 1:4) plot(brain.wpc, level=i, power=.2)
```

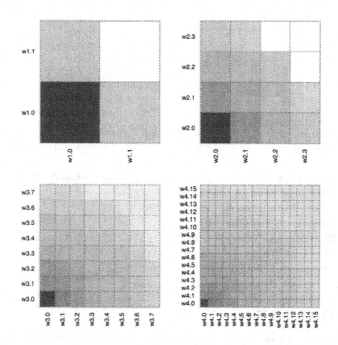

FIGURE 9.9. Wavelet packet cost table for the **brain** image, with level 1 costs to level 4 costs plotted as images.

The 2-D best basis algorithm searches the 2-D cost table just as in the 1-D case. Produce an EDA plot of the wavelet packet cost table as follows:

```
> eda.plot(brain.wpc)
```

The EDA plot is shown in figure 9.10. A dot chart ranks the "best basis," the DWT, and the level 0 to level 6 WPTs by entropy. The DWT is virtually as good as the best basis, and is better than the block WPTs.

Compute a cosine packet cost table as follows:

```
> brain.cpc <- cp.costs.2d(brain - mean(brain),
+    n.level=6)
> brain.cpc
```

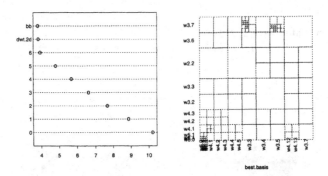

FIGURE 9.10. EDA plot for the `brain` image. Left: a dot chart ranking the "best basis", the DWT, and the level 0 to level 6 WPTs. Right: a grid of the best basis.

```
Cosine packet cost vector for brain - mean(brain)
Cost function: threshold
Image Dimensions:  256 by 256
Number of Levels:  6
Horizontal Taper Function:  poly2
Vertical Taper Function:  poly2
Horizontal Boundary Rule:  periodic
Vertical Boundary Rule:  periodic
Horizontal DCT-type:  2
Vertical DCT-type:  2
```

By default, a cosine packet cost table uses the **threshold** cost measure, which counts the number of coefficients above a certain threshold. The threshold is set to the upper quartile (75%) of the absolute value of the DCT of the entire image.

The best basis from the table is computed with the **best.basis** function:

```
> brain.cpbb <- best.basis(brain.cpc)
> length(brain.cpbb)
[1] 187
```

This computes only the *names* of the crystals corresponding to the best basis. There are 187 different crystals in the best basis for the brain image. Only the names are computed because the **brain.cpc** object contains only the table of costs, and not the cosine packet coefficients. To actually compute the best basis, you need to supply the original image to the **best.basis** function using the argument **data**. An example of this is provided in the next section on page 193.

Use the function `pgrid.plot` to superimpose the blocking for a cosine packet best basis on the original image:

```
> image(brain)
> pgrid.plot(brain.cpbb, add=T, dim=dim(brain), axes=F)
```

This results in figure 9.11. The best basis selects smaller blocks to handle finer features and edges and larger blocks to represent relatively homogeneous regions.

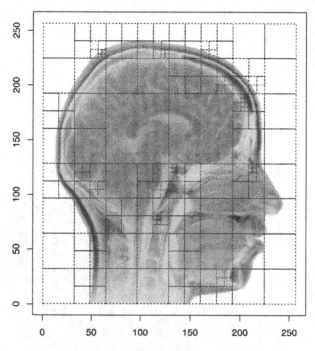

FIGURE 9.11. Cosine packet best basis for the **brain** image.

Note: The **entropy** cost measure is not very useful for 2-D cosine packet analysis; the mean term of the block can swamp the rest of the information in the block. That is why the **threshold** cost function is used as the default for 2-D cosine packets. An alternative is the **cpentropy** cost measure, which omits the mean terms from the entropy calculations.

9.4   Analysis of a Digital Fingerprint Image

One of the great success stories of wavelets lies in their use to compress digital fingerprints [Cri93]. The S+WAVELETS built-in object **fingerprint** is a cropped 512×512 digital fingerprint image. While you can work with the original image, in order to speed the computations, you might want to reduce this to a 256×256 image. One way to do this is to take 2×2 block averages. This is done in S+WAVELETS by taking the **s1-s1** component from the Haar wavelet transform. Reduce the **fingerprint** image and plot the resulting image with

```
> finger0 <- dwt.2d(fingerprint, wavelet="haar",
+     n.level=1)[["s1-s1"]]
> finger0 <- as.matrix(finger0)
> image(finger0)
```

The 256×256 **finger0** image is shown in figure 9.12.

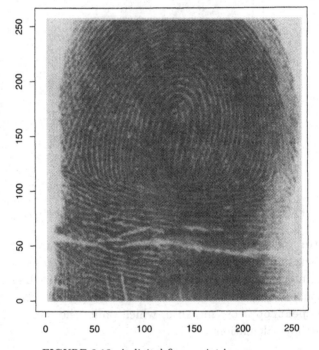

FIGURE 9.12. A digital fingerprint image.

9.4.1 Proposed FBI Standard for Compression

The proposed fingerprint compression standard is based on a specific wavelet packet transform with 64 crystals or subbands. The crystal names for this transform are available in the data object fingerprint.basis. Subtract the mean from the finger0 image and compute the 2-D DWT and WPT based on the proposed FBI standard:

```
> finger0.mean <- mean(finger0)
> finger1 <- finger0 - finger0.mean
> finger.dwt <- dwt.2d(finger1, n.level=5)
> finger.wpt <- wpt.2d(finger1,
+   crystal.names=fingerprint.basis)
```

Next plot the resulting DWT to produce figure 9.13:

```
> plot(finger.dwt, power=0)
```

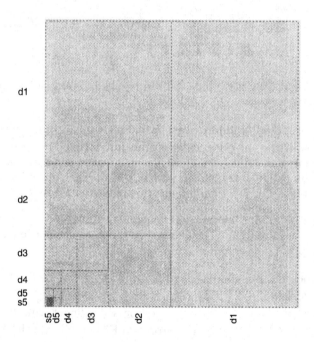

FIGURE 9.13. Plot of the discrete wavelet transform of the fingerprint image.

Now plot the WPT shown in figure 9.14:

```
> plot(finger.wpt, power=0)
```

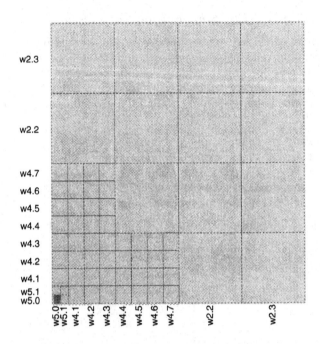

FIGURE 9.14. Plot of the proposed FBI standard wavelet packet transform of the fingerprint image.

The proposed standard wavelet packet transform differs from the DWT, splitting the finer scale crystals into either 4 or 16 subcrystals.

The current JPEG standard for still image compression is based on the block DCT transform with 8×8 blocks. Compute and plot a level 5 block DCT for the fingerprint image as follows:

```
> finger.dct <- cpt.2d(finger1, n.level=5,
+                      boundary="periodic", taper="boxcar")
> plot(finger.dct)
```

The block DCT is displayed in figure 9.15. A level 5 DCT produces 8×8 blocks $(256/2^5 = 8)$.

Which transform is best for compressing fingerprint images: the DWT, the WPT based on the proposed FBI standard ("FBI-WPT"), or the traditional block DCT? To answer this question, we can compare the mean square error (MSE) of the reconstructed images based on the largest coefficients. Define the energy concentration function

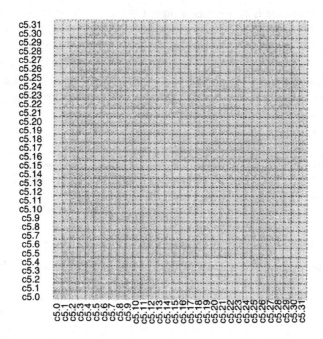

FIGURE 9.15. Level 5 block DCT of the fingerprint image.

for a vector $\mathbf{x} = (x_1, x_2, \ldots, x_n)'$ by

$$E_x(K) = \frac{\sum_{i=1}^{K} x_{(i)}^2}{\sum_{i=1}^{n} x_i^2}$$

where $x_{(i)}$ is the ith largest value in $\mathbf{x}$. Since these are orthogonal transforms, the MSE of the reconstructed image based on the largest K coefficients is given by $E_x(K)$.

We will compute $E_x(K)$ for the DWT, the FBI-WPT, and the block DCT. First sort the squared transform coefficients from largest to smallest:

```
> dwt.sort <- rev(sort(unclass(finger.dwt)^2))
> wpt.sort <- rev(sort(unclass(finger.wpt)^2))
> dct.sort <- rev(sort(unclass(finger.dct)^2))
```

We need to unclass the objects in this case in order to treat the coefficients as a single long vector, and not as a collection of crystals. Next compute the energy functions by applying the cumsum operator and dividing by the total energy of the image:

```
> total.energy <- sum(finger1^2)
> dwt.energy <- cumsum(dwt.sort)/total.energy
> wpt.energy <- cumsum(wpt.sort)/total.energy
> dct.energy <- cumsum(dct.sort)/total.energy
```

We are now ready to plot the percentage MSE as a function of the compression ratio (the original number of pixels in the image divided by the number of coefficients retained):

```
> N <- length(dct.energy)
> ind <- seq(from=N/100, to=N, length=500)
> plot(N/ind, 100*(1-dct.energy[ind]), type="l",
+            xlab="Compression Ratio", ylab="MSE (%)")
> lines(N/ind, 100*(1-dwt.energy[ind]), lty=2)
> lines(N/ind, 100*(1-wpt.energy[ind]), lty=3)
> abline(v=20,lty=2)
> legend(70, 4, c("DCT", "DWT", "FBI"), lty=1:3)
```

The resulting plot is shown in figure 9.16. The proposed FBI stan-

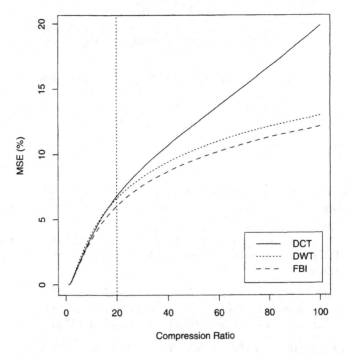

FIGURE 9.16. Comparison of the energy concentration for the DWT, the FBI standard, and the level 5 block DCT. The vertical line marks the compression ratio of 20:1.

dard WPT has uniformly smaller MSE than the DWT and the block CPT. For very high compression ratios, the block CPT has much higher MSE.

The ultimate proof in image compression lies in the visual quality of the reconstructions. Wipe out all but the largest 5% of the coefficients for the three transforms as follows:

```
> dct.thresh <- sqrt(dct.sort[floor(N/20)])
> wpt.thresh <- sqrt(wpt.sort[floor(N/20)])
> dwt.thresh <- sqrt(dwt.sort[floor(N/20)])
> finger.dct[abs(finger.dct) < dct.thresh] <- 0
> finger.dwt[abs(finger.dwt) < dwt.thresh] <- 0
> finger.wpt[abs(finger.wpt) < wpt.thresh] <- 0
```

Reconstruct the FBI-WPT and compare with the original finger1 image, remembering to add in the mean of the original image, given by finger0.mean:

```
> fing.wpt.hat <- reconstruct(finger.wpt)+finger0.mean
> image(fing.wpt.hat)
```

The images are shown in figure 9.17. Using just 5% of the coefficients, the reconstructed image does a good job at representing the fingerprint image.

Warning: To fully see the differences between the original and compressed fingerprint images, you should reproduce the plots of the fingerprint image on a high-resolution printer or monitor.

Reconstruct and plot the DWT representations to produce figure 9.18:

```
> fing.dwt.hat <- reconstruct(finger.dwt)+finger0.mean
> image(fing.dwt.hat)
```

Finally, reconstruct and plot the DCT representations to produce figure 9.19:

```
> fing.dct.hat <- reconstruct(finger.dct)+finger0.mean
> image(fing.dct.hat)
```

These DWT and DCT reconstructions are perceptually worse than the FBI-WPT reconstruction, with visibly higher distortion in parts of the image.

Compare the residual image from the FBI-WPT and the block DCT reconstructions with the following:

```
> par(mfrow=c(1,2))
> image(abs(finger0-fing.wpt.hat))
> image(abs(finger0-fing.dct.hat))
```

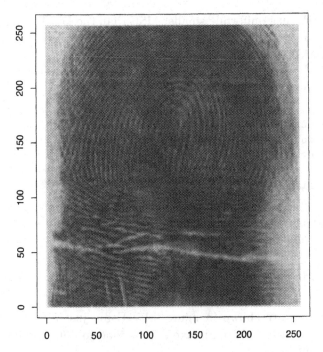

FIGURE 9.17. The reconstructed fingerprint image using the top 5% of the co-efficients from the FBI wavelet packet transform.

Figure 9.20 displays the FBI-WPT residual image (left) and the block DCT residual image (right). The whorls in the block DCT residual image are visibly more pronounced, indicating that the DCT is less effective in capturing the peaks and valleys of the fingerprint image.

Note: The actual FBI standard uses the biorthogonal wavelet **vs1**. Orthogonal wavelets are used in this section to allow easy comparison of the MSE as a function of compression ratio (a biorthogonal transform is not energy preserving since the coefficients are correlated). Also, the FBI standard does not literally set 95% of the coefficients to zero, but instead uses *scalar quantization* to compress floating point numbers into codewords of a fixed bit rate. See the proposed standard [Cri93] for details.

9.4.2   Best Basis Analysis of Fingerprints

We continue our analysis of the digital fingerprint image by computing a wavelet packet and cosine packet cost table and finding the

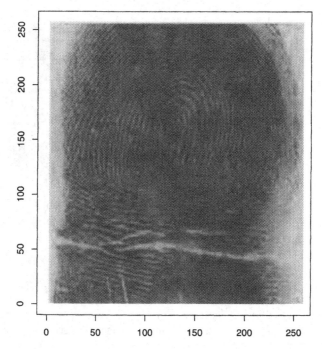

FIGURE 9.18. The reconstructed fingerprint image using the top 5% of the co-efficients from the DWT.

"best basis" from these tables. First we compute a wavelet packet table with 5 levels and the corresponding EDA plot:

```
> finger.wpc <- wp.costs.2d(finger1, n.level=5)
> eda.plot(finger.wpc)
```

This produces figure 9.21. The threshold cost function is used here instead of the usual entropy cost function. The original image, level 1, and level 2 transforms all lead to much higher entropy. The best basis has significantly lower entropy than the remaining transforms.

Now compute a cosine packet table and EDA plot for the finger-print image:

```
> finger.cpc <- cp.costs.2d(finger1, n.level=5,
+     boundary="periodic")
> eda.plot(finger.cpc)
```

The EDA plot is displayed in figure 9.22. The best basis transform gives a dramatic reduction in cost over the other transforms. A non-uniform blocking affords significant advantages for data compression over the uniform blocking of the JPEG method.

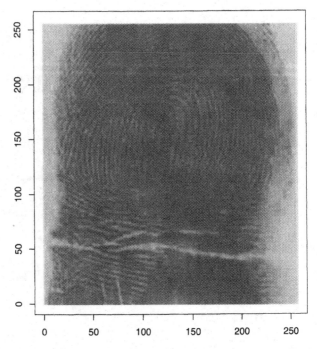

FIGURE 9.19. The reconstructed fingerprint image using the top 5% of the co-efficients from the DCT.

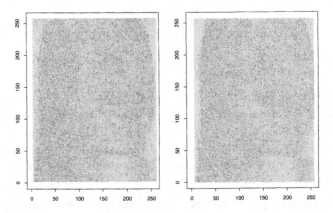

FIGURE 9.20. The residual images for the reconstructed fingerprint images using the top 5% of the coefficients. Left: from the FBI-WPT. Right: from the block DCT.

Compute the best basis transforms using the wavelet packet and cosine packet cost tables as follows:

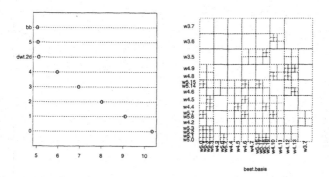

FIGURE 9.21. EDA plot of a wavelet packet cost table. Left: comparing the best basis to the levels and the DWT. Right: the best basis grid.

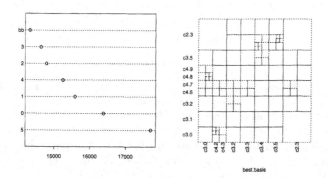

FIGURE 9.22. EDA Plot of a cosine packet cost table. The left plot compares the best basis to the levels and the right plot shows the best basis grid.

```
> finger.wpbb <- best.basis(finger.wpc, data=finger1)
> finger.cpbb <- best.basis(finger.cpc, data=finger1)
```

The argument `data=finger1` causes the actual transform to be computed, and not just the names of the crystals. You need to supply the original image because the cost tables contain only costs, and not transform coefficients.

Now compute the energy functions $E(K)$:

```
> wpbb.sort <- rev(sort(unclass(finger.wpbb)^2))
> cpbb.sort <- rev(sort(unclass(finger.cpbb)^2))
> wpbb.energy <- cumsum(wpbb.sort)/total.energy
> cpbb.energy <- cumsum(cpbb.sort)/total.energy
```

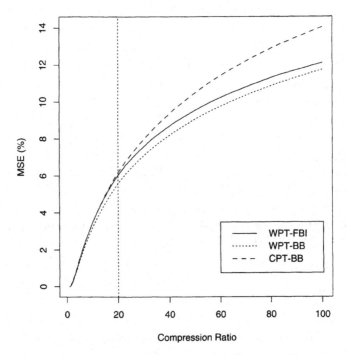

FIGURE 9.23. Comparison of the energy concentration for the FBI standard WPT (WPT-FBI), the wavelet packet best basis transform (WPT-BB), and the cosine packet best basis transform (CPT-BB).

These are plotted these as in section 9.4.1 to compare the MSE of the wavelet packet best basis (WPT-BB), cosine packet best basis (CPT-BB), and the proposed standard (WPT-FBI):

```
> N <- length(wpt.energy)
> ind <- seq(from=N/100, to=N, length=500)
> plot(N/ind, 100*(1-wpt.energy[ind]), type="l",
+      xlab="Compression Ratio", ylab="MSE (%)",
+      ylim = range(100*(1-cpbb.energy[ind])))
> lines(N/ind, 100*(1-wpbb.energy[ind]), lty=2)
> lines(N/ind, 100*(1-cpbb.energy[ind]), lty=3)
> abline(v=20,lty=2)
> legend(60, 3.5, c("WPT-FBI", "WPT-BB", "CPT-BB"),
+        lty=1:3)
```

This results in figure 9.23. The wavelet packet best basis transform has the lowest MSE uniformly across all compression ratios. It is not surprising that WPT-BB leads to better compression than WPT-FBI (in terms of MSE), since WPT-BB has been optimized for this image whereas WPT-FBI is a standard meant to work for all images. At lower compression ratios, the WPT-FBI transform matches the MSE of the cosine packet best basis transform. At high compression rations, WPT-FBI has much lower MSE than CPT-BB.

Next, set all but the largest 5% of the coefficients to zero for the wavelet packet best basis transform:

```
> wpbb.thresh <- sqrt(wpbb.sort[floor(N/20)])
> finger.wpbb[abs(finger.wpbb) < wpbb.thresh] <- 0
```

Now reconstruct from the best basis transform and plot the reconstruction shown in figure 9.24:

```
> image(reconstruct(finger.wpbb)+finger0.mean)
```

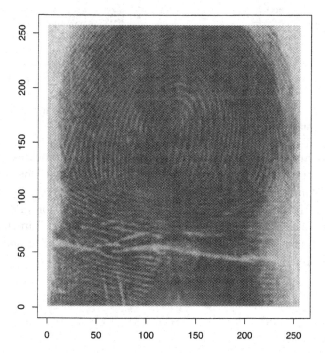

FIGURE 9.24. The reconstructed fingerprint image using the top 5% of the coefficients from the best basis WPT.

The best basis WPT is perceptually superior than the the FBI wavelet packet transform (compare figure 9.17).

Finally, plot the location of the top 5% of the coefficients for the

FBI wavelet packet transform and the best basis WPT:

```
> par(mfrow=c(1,2))
> plot(abs(finger.wpt)>0)
> plot(abs(finger.wpbb)>0)
```

The result is shown in figure 9.25. The large coefficients are located

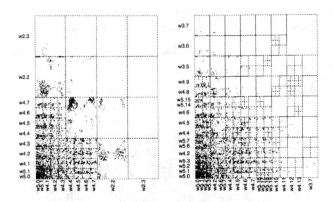

FIGURE 9.25. Location of the top 5% of the coefficients in the FBI wavelet packet transform (left) and the best basis WPT (right).

in roughly the same crystals despite the different blocking factors.

The crystal w2.0-w2.2 in the FBI-WPT has numerous large coefficients in the lower left corner. This crystal is important in capturing the "horizontal whorls" in the lower left hand portion of the image. The crystal w2.2-w2.0 captures the "vertical whorls" in the middle of the image. The crystal w2.1-w2.1 captures the "diagonal whorls" in the upper corners.

10

Matching Pursuit and Molecules

The matching pursuit decomposition of Mallat and Zhang [MZ93] is an alternative to wavelet packet and cosine packet analysis which can provide greater flexibility in analysis of signals. In this chapter, you will learn about the following topics:

- Matching pursuit decompositions and the matching pursuit algorithm (section 10.1).

- Computing a matching pursuit decomposition (section 10.2).

- Molecule and atom objects (section 10.3).

- How the matching pursuit compares with the DWT and with the wavelet packet and cosine packet transforms (section 10.4).

10.1 Matching Pursuit Decompositions

Matching pursuit is an algorithm developed by Mallat and Zhang [MZ93] in which a signal is decomposed as a sum of "atomic waveforms" $g_\gamma(t)$. The matching pursuit algorithm can be used with many different collections of atomic waveforms $\{g_\gamma(t) : \gamma \in \Gamma\}$, which Mallat and Zhang call "dictionaries." Two examples of dictionaries are wavelet packet tables and cosine packet tables. In contrast to the discrete wavelet transform and best basis transforms, matching pursuit decompositions are not orthogonal.

The matching pursuit approximation to a signal $f(t)$ as a sum of N atomic waveforms $g_{\gamma_1}(t)$, $g_{\gamma_2}(t)$, ..., $g_{\gamma_N}(t)$ from a dictionary Γ is given by

$$f(t) = \sum_{n=1}^{N} h_n g_{\gamma_n}(t) + R_N(t). \qquad (10.1)$$

The h_n are the *matching pursuit coefficients* and $R_N(t)$ is the *residual*.

The matching pursuit algorithm proceeds by successively finding and removing atomic waveforms from a signal $f(t)$. An approximation with N atoms is obtained as follows:

[0] Initialize $R_0(t) = f(t)$ and $n = 1$.

[1] For all atomic waveforms $g_\gamma, \gamma \in \Gamma$, compute the projections

$$c_{\gamma,n} = \int R_{n-1}(t) g_\gamma(t) dt. \qquad (10.2)$$

Note that the coefficients $c_{\gamma,n}$ depend on the iteration number n.

[2] Find the index n of the atomic waveform γ_n with the maximum projection:

$$\begin{aligned} \gamma_n &= \text{argmin}_{\gamma \in \Gamma} \| R_{n-1}(t) - c_{\gamma,n} g_\gamma(t) \| \\ &= \text{argmax}_{\gamma \in \Gamma} |c_{\gamma,n}|. \end{aligned}$$

The second equality follows from an energy conservation equation (see below). The nth matching pursuit atom is given by g_{γ_n}.

[3] Compute the nth residual $R_n(t)$ as

$$R_n(t) = R_{n-1}(t) - h_n g_{\gamma_n}(t). \qquad (10.3)$$

[4] Set $n = n + 1$. If $n \leq N$, then go to step [1].

[5] After N iterations, perform the *back-projection* step. Let $\mathbf{P}_N$ the orthogonal projector onto $g_{\gamma_1}(t)$, $g_{\gamma_2}(t)$, ..., $g_{\gamma_N}(t)$. Compute

$$\mathbf{P}_N R_N(t) = \sum_{n=1}^{N} x_n g_{\gamma_1}(t).$$

The matching pursuit coefficients are given by $h_n \equiv c_{\gamma_n, n} + x_n$.

Typically, all of the atomic waveforms in the dictionary will have unit norm, so $\| g_\gamma(t) \| = 1$ for all $\gamma \in \Gamma$ (as with wavelet packet and cosine packet dictionaries). For these dictionaries, since g_{γ_n} is orthogonal to the residual $R_n(t)$;

$$\| R_{n-1}(t) \|^2 = h_n^2 + \| R_n(t) \|^2 . \tag{10.4}$$

This leads to the energy conservation equation

$$\| f(t) \|^2 = \sum_{n=1}^{N} h_n^2 + \| R_N(t) \|^2 \tag{10.5}$$

even though the atomic waveforms g_{γ_n} may not be orthogonal.

The matching pursuit algorithm is very similar to the algorithm used by Friedman and Stuetzle [FS81] for project pursuit regression. You can perform a project pursuit regression in S-PLUS with the function **ppreg**. For more information about projection pursuit regression, refer to the chapter on Regression and Smoothing for Continuous Response Data (for UNIX customers, this chapter is in the *S-PLUS Guide to Statistical and Mathematical Analysis*; for Windows customers, it is in the *S-PLUS for Windows Version 3.2 Supplement*).

S+WAVELETS offers two types of dictionaries for matching pursuit: wavelet packet tables and cosine packet tables. A matching pursuit approximation using wavelet packet tables is given by

$$f(t) = \sum_{j} \sum_{b} \sum_{k} w_{j,b,k} W_{j,b,k}(t) + R_N(t) \tag{10.6}$$

where j is the level, b is the oscillation, and k is the shift. Similarly, the matching pursuit approximation using cosine packet tables is given by

$$f(t) = \sum_{j} \sum_{b} \sum_{k} c_{j,b,k} C_{j,b,k}(t) + R_N(t) \tag{10.7}$$

where j is the level, b is the block, and k is the frequency. See chapters 7 and 8 for more information about wavelet packet and cosine packet functions.

Note: Mallat and Zhang also use dictionaries of Gabor functions, which are the product of a window or taper function and a complex exponential function. Gabor dictionaries provide a richer collection of atomic waveforms which are located on a much finer grid in time-frequency space than wavelet packet and cosine packet tables. See Mallat and Zhang [MZ93] for details.

10.2 Computing a Matching Pursuit Transform

First, create and plot the quadratic chirp signal shown in figure 10.1 with:

```
> chirp2 <- make.signal("quadchirp", n=512)
> ts.plot(chirp2)
```

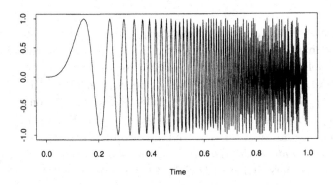

FIGURE 10.1. A quadratric chirp signal.

The function `matching.pursuit` computes a wavelet packet or cosine packet matching pursuit transform. Obtain and plot a wavelet packet matching pursuit transform of the quadratic chirp signal by:

```
> chirp2.mp <- matching.pursuit(chirp2, type="wp",
+          n.atom=50)
> plot(chirp2.mp)
```

The argument `type="wp"` creates a wavelet packet matching pursuit transform. To obtain a cosine packet matching pursuit transform, use `type="cp"`. The `plot` function displays the coefficients of the matching pursuit decomposition as a wavelet packet table, as illustrated in figure 10.2. The vertical scale of all coefficients is the same,

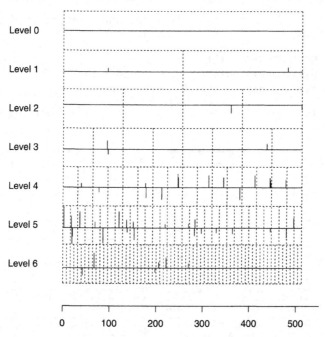

FIGURE 10.2. A plot of the coefficients in the wavelet packet table for a matching pursuit decomposition with 50 atoms.

so the length of the line indicates the magnitude of the coefficient. Note that the coefficients are spread throughout the table.

Here is a summary of the matching pursuit decompositions:

```
> summary(chirp2.mp)
          #    Min      1Q Median     3Q     Max Energy %
level.0   0  0.000   0.000  0.000  0.000   0.000   0.000
level.1   2  0.810   0.842  0.874  0.906   0.938   0.005
level.2   2 -1.769  -1.572 -1.375 -1.178  -0.981   0.013
level.3   3 -1.090   0.125  1.340  1.653   1.966   0.036
level.4  15 -2.719  -0.114  1.530  2.210   3.027   0.304
level.5  22 -3.570  -1.620 -0.006  1.866   4.751   0.571
level.6   6 -1.580  -0.434  1.137  2.051   3.315   0.071
```

There is a non-zero coefficient in level 0. This is not possible for orthogonal wavelet packet transforms of the type discussed in chapter 7 (except for the trivial basis).

Now extract the top 50 atoms from a DWT of the quadratic chirp signal with the `top.atoms` function:

```
> chirp2.dwt <- top.atoms(dwt(chirp2), n.atom=50)
```

Plot the top 25 atoms of the DWT and matching pursuit decompositions with:

```
> par(mfrow=c(2,1))
> plot(decompose(chirp2.mp), n.top=25)
> plot(decompose(chirp2.dwt), n.top=25)
```

The resulting plot is shown in figure 10.3.   The matching pursuit

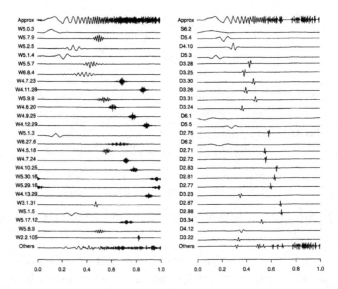

FIGURE 10.3. The top 25 atoms quadratic chirp decomposition. Left: matching pursuit decomposition. Right: DWT decomposition.

wavelet packet approximation of the quadratic chirp based on 50 atoms is far superior to the DWT approximation. Matching pursuit uses a much richer collection of atomic waveforms to represent the oscillatory nature of the chirp.

Warning: For both wavelet packets and cosine packets, the function `matching.pursuit` applies only to sample sizes divisible by 2^J, where J is the maximum level in the wavelet packet table. In addition, for wavelet packets, you are restricted to orthogonal wavelets with the `periodic` boundary condition.

10.3   Molecule and Atom Objects

The `matching.pursuit` and `top.atoms` functions create a *molecule* object. Look at the `chirp2.mp` wavelet packet molecule:

```
> chirp2.mp
Wavelet molecule from signal 'chirp2'
Produced by match pursuit method
Wavelet: s8
Length of series: 512
Number of levels: 6
Boundary correction rule: periodic
Top 32 atoms: (total 50 atoms)
    w5.0.3   w5.7.9   w5.2.5    w5.1.4    w5.5.7
   4.750607 3.62028 3.593282 -3.569891 -3.372731
   w6.8.4   w4.7.23  w4.11.28   w5.9.9    w4.6.20
   3.3153 3.026505 -2.719221 -2.665639 -2.651898
   w4.9.25 w4.12.29   w5.1.3   w6.27.6    w4.5.18
   2.649454 2.607644 2.581447 2.305817 -2.267115
   w4.7.24 w4.10.25 w5.30.16 w5.29.16 w4.13.29
   2.212637 2.207996 2.178701  -2.1718 2.096043
   w3.1.31    w5.1.5 w5.17.12   w5.8.9   w2.2.105
   1.965835 -1.899671 1.890246 1.794955 -1.768879
   w5.17.11 w4.14.30   w6.5.3 w4.13.30   w3.6.54
   -1.755424 1.692814 -1.58017 1.529603 1.339612
    w5.9.7    w5.4.6
   1.332823 1.308194
```

A wavelet packet molecule is a vector of wavelet packet coefficients $w_{j,b,k}$. In contrast to a crystal, the coefficients are not arranged on a lattice, and may take any order. When created by `matching.pursuit` and `top.atoms`, a molecule is ordered by size, so that the largest coefficients are first.

As with crystals, you can apply a number of generic functions to molecules, such as `plot`, `summary`, `eda.plot`, and `reconstruct`. Use the `[` operator to subset molecules to create other molecules and use the assignment operator `<-` to assign portions of the molecule to a numeric value.

Using the `[[` operator, which extracts a signal element from an object, you can create an *atom*. Wavelet packet atoms are just scaled and translated wavelet packet functions. These atoms are essentially molecules of length one, and consist of a coefficient $w_{j,b,k}$ associated with a wavelet packet function $W_{j,b,k}(t)$ (or cosine packet function). Extract the atom $w_{4,11,28} W_{4,11,28}(t)$ with:

```
> atom <- chirp2.mp[["w4.11.28"]]
> atom
s8 Wavelet Atom: w4.11.28
Coefficient: -2.719221
```

You can plot this atom with

```
> plot(atom)
```

The result is shown in figure 10.4.

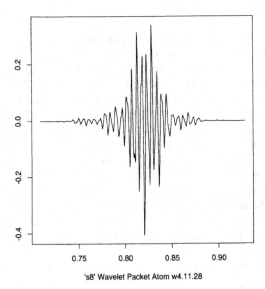

's8' Wavelet Packet Atom w4.11.28

FIGURE 10.4. A plot of the wavelet packet atom $w_{4,11,28} W_{4,11,28}(t)$.

10.4   Comparison with DWT and Best Basis

In this section, we will compare matching pursuit to the DWT and the wavelet packet best basis transform. Create and plot the complex signal, shown in figure 10.5, with:

```
> signal <- make.signal("mishmash3", n=512)
> ts.plot(signal)
```

This signal is the sum of a quadratric chirp, a singularity at $t \approx .37$, and two sinusoids with frequencies .33 and .02. The rts function is used to convert the sinusoid signals into "time series objects" from $t = 0$ to $t = 1$.

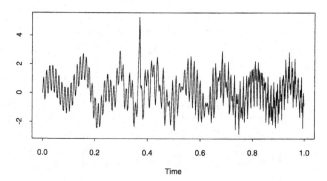

FIGURE 10.5. A complex signal composed of a quadratic chirp, two sinusoids of different with frequencies .33 and .02, and a singularity.

Now compute the top 100 atoms for the matching pursuit, the wavelet packet best basis, and the discrete wavelet transforms:

```
> signal.mp <- matching.pursuit(signal, type="wp",
+       n.atoms=100)
> signal.bb <- top.atoms(best.basis(wp.table(signal)),
+       n.atoms=100)
> signal.dwt <- top.atoms(dwt(signal), n.atoms=100)
```

Compare the location of the atoms for the best basis and matching pursuit transforms with:

```
> par(mfrow=c(1,2))
> plot(signal.mp)
> plot(signal.bb)
```

The resulting plot is shown in figure 10.6.   The matching pursuit decomposition has coefficients spread throughout the table. By contrast, the best basis transform coefficients are located in just a few crystals. Of particular note is the coefficient in level 1 for the matching pursuit decomposition. This coefficient is associated with the singularity located at $t \approx .37$. This shows an advantage of the matching pursuit algorithm: it is able to efficiently represent local features in time and frequency.

Another way to compare the matching pursuit and best basis transforms is to look at the proportion of energy by crystal. You can plot the intensity of the energy as follows:

```
> par(mfrow=c(1,2))
> plot(pcosts(as.ptable(signal.mp),cost.fun="energy",
+       scale=1), power=.5)
> plot(pcosts(as.ptable(signal.bb),cost.fun="energy",
```

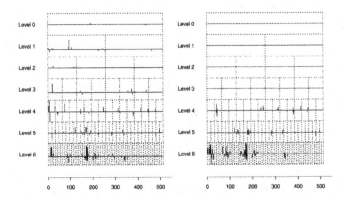

FIGURE 10.6. The top 100 atoms for the complex signal. Left: the matching pursuit transform. Right: best wavelet packet basis transform.

```
+        scale=1), power=.5)
```

The result is shown in figure 10.7 with dark rectangles corresponding to high energy crystals. In the above expression, the `as.ptable` function is used to convert the molecule to a wavelet packet table and the `pcosts` function is used to compute the energy.

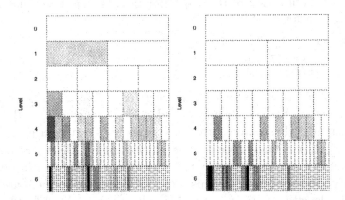

FIGURE 10.7. The proportion of energy by crystal for the top 100 atoms. Left: the matching pursuit transform. Right: the best wavelet packet basis transform.

Compare the energy concentration function $E_x(K)$, defined in (4.3), for the matching pursuit decomposition, DWT, and best basis transform as follows:

```
> # tot.e = total energy
> tot.e <- sum(signal^2)
```

```
> energy.mp <- cumsum(as.vector(signal.mp^2))/tot.e
> energy.dwt <- cumsum(as.vector(signal.dwt^2))/tot.e
> energy.bb <- cumsum(as.vector(signal.bb^2))/tot.e
> plot(energy.mp, type="l", ylim=c(0,1))
> lines(energy.dwt, lty=2)
> lines(energy.bb, lty=3)
> legend(40,.4, lty=1:3,
+   legend=c("matching pursuit","DWT","best basis"))
```

These expressions produce figure 10.8.   The matching pursuit de-

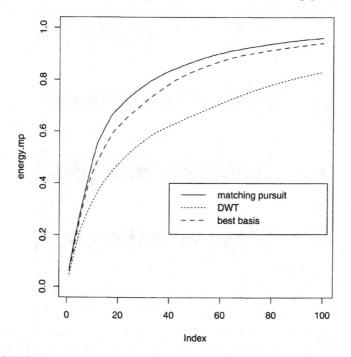

FIGURE 10.8. The energy concentration function for the top 100 atoms.

composition uniformly captures the most energy because matching pursuit is a "greedy algorithm," and at each iteration finds the atom with the most energy. The best basis results from a global optimization.

Figure 10.9 shows the residuals for each decomposition, computed as follows:

```
> signal.stack <- list(data=signal,
+        signal.mp=signal-reconstruct(signal.mp),
+        signal.dwt=signal-reconstruct(signal.dwt),
```

```
+          signal.bb=signal-reconstruct(signal.bb))
> stack.plot(signal.stack, zerocenter=T, zeroline=T,
+     same.scale=2:4)
```

The argument `same.scale=2:4` forces the residuals plotted with the same vertical scale while the signal is plotted with its own vertical scale. The matching pursuit approximation leads to the smallest residuals.

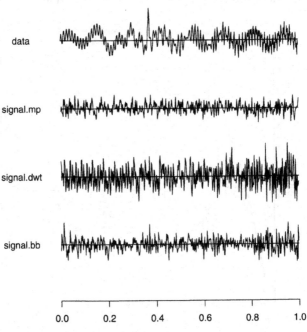

FIGURE 10.9. Comparison of residuals for three decompositions of quadratic chirp signal. Top: original signal. Second: the residual term based on the top 100 atoms for matching pursuit decomposition. Third: residuals from DWT. Bottom: residuals from best basis.

Time-frequency plots for the matching pursuit and best basis decompositions, shown in figure 10.10, can be computed as follows:

```
> par(mfrow=c(1,2))
> time.freq.plot(signal.mp)
> time.freq.plot(signal.bb)
```

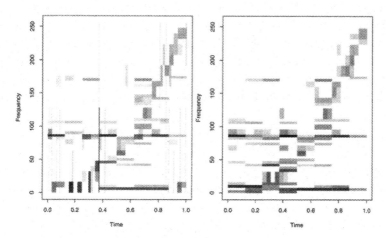

FIGURE 10.10. Time-frequency plots of the quadratic chirp. Left: using the matching pursuit decomposition. Right: using best basis decomposition.

The matching pursuit time-frequency plot does a better job of identifying the singularity while the best basis decomposition more clearly identifies the sinusoids and the quadratic chirp.

11

Variations on Wavelet Analysis

In this chapter, we will explore some variations on wavelet analysis, including the following topics:

- Non-decimated wavelet transforms, which provide better spatial resolution (section 11.1).

- A "robust" wavelet decomposition which is resistant to outliers in your signal (section 11.2).

- How to create and use your own wavelets for analysis (section 11.3).

11.1 Non-Decimated Wavelets

The non-decimated discrete wavelet transform, also known as the stationary wavelet transform or translation invariant wavelet transform, is a non-orthogonal variant of the classical DWT. In the non-decimated DWT, starting with n sample signal values one ends up with $(J + 1) \times n$ coefficients. Unlike the classical DWT, which has fewer coefficients at coarser scales, each scale for the non-decimated DWT has n coefficients. The non-decimated DWT is *over-sampled* at coarse scales.

Unlike the orthogonal DWT, the non-decimated DWT is *translation invariant*. In other words, shifts in the signal do not change the pattern of coefficients for the non-decimated DWT. This is a key property which leads to advantages in certain problems and enhances the visual displays. In particular, using the non-decimated DWT in place of the DWT in the WaveShrink algorithm leads to both better prediction and fewer artifacts, such as Gibbs phenomena near discontinuities [CD95]. See [MZ92, She92, BG94, LGO+95, NS95, PM95] for further discussion of the advantages and properties of the non-decimated DWT.

A non-decimated discrete wavelet transform is obtained using the S+WAVELETS nd.dwt function. Compute a non-decimated DWT for the glint signal as follows:

```
> glint.nd.dwt <- nd.dwt(glint)
> glint.nd.dwt
Non-Decimate Discrete Wavelet Transform of glint
Wavelet: s8
Length of series: 512
Number of levels: 6
Boundary correction rule: periodic
Crystals: s6 d6 d5 d4 d3 d2 d1
```

Next plot the glint.nd.dwt object:

```
> plot(glint.nd.dwt)
```

The non-decimated wavelet transform is pictured in figure 11.1. In contrast to the classical wavelet transform, the non-decimated DWT provides better spatial resolution, particularly at coarser scales; compare with figure 4.2.

Even though the non-decimated wavelet transform is not orthogonal, you can still use the reconstruct function to invert the transform to obtain the original series.

11.1.1   Wavelet Shrinkage with Non-Decimated Wavelets

In chapter 6, you learned about using wavelet shrinkage as a nonparametric tool for de-noising signals. Instead of using an orthogonal discrete wavelet transform, you can de-noise your data using the non-decimated wavelet transform.

As an example, we create a noisy jumpsine signal and compare "classical" wavelet shrinkage based on the orthogonal haar wavelet transform with wavelet shrinkage resulting from a non-decimated

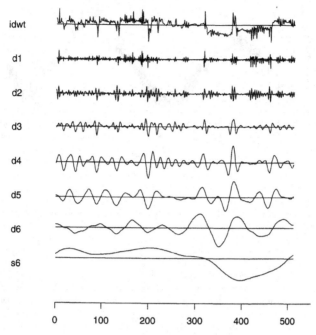

FIGURE 11.1. Non-decimated DWT for the **glint** signal.

haar wavelet transform. The result is shown in figure 11.2:

```
> par(mfrow=c(2,1))
> noisy.jumpsine <- make.signal("jumpsine", snr=4)
> ws.jumpsine <- waveshrink(noisy.jumpsine,
+    wavelet="haar")
> nd.jumpsine <- nd.dwt(noisy.jumpsine, wavelet="haar")
> ws.nd.jumpsine <- waveshrink(nd.jumpsine)
> plot(noisy.jumpsine, type="l")
> plot(ws.jumpsine, type="l")
> lines(ws.nd.jumpsine, lty=2)
```

Note: You can also compute the non-decimated WaveShrink estimate directly, as follows:

```
> waveshrink(noisy.jumpsine, wavelet="haar", decimate=F)
```

The orthogonal **haar** wavelet estimate is very blocky, corresponding to the discontinuous nature of the **haar** wavelet. By contrast, the non-decimated **haar** wavelet estimate is smooth due to the "spatial averaging" in the reconstruction.   There is, however, a cost to smoothness: the non-decimated estimate blurs the discontinuities at

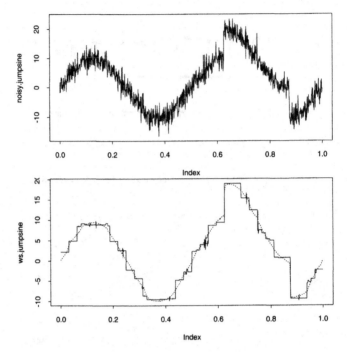

FIGURE 11.2. WaveShrink with non-decimated wavelets. Top: the noisy jump-sine signal. Bottom (solid line): smooth produced by wavelet shrinkage using the orthogonal DWT. Bottom (dashed line): smooth produced by wavelet shrinkage using the non-decimated DWT. The **haar** wavelet is used in this example.

the two jumps.

Obtain an EDA plot for the non-decimated estimate as follows:

```
> eda.plot(ws.nd.jumpsine)
```

This is displayed in figure 11.3. The non-decimated wavelet estimate appears to do a good job of separating the signal from the data in this example.

Note: You can see that the only fine scale detail coefficients which were not set to zero are those associated with the jumps. To avoid the blurring of the jumps, it is desirable to do even less shrinkage at the jump points. One approach which achieves this aim is to use the coarse scale coefficients to help determine which fine scale coefficients should be retained; see [MH92].

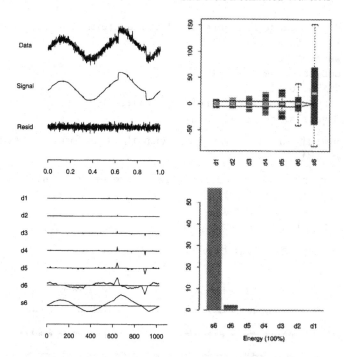

FIGURE 11.3. EDA plot for wavelet shrinkage of the noisy jumpsine signal using the the non-decimated DWT.

11.1.2   The "à trous" Wavelet Transform

An alternative non-decimated wavelet transform is given by the "à trous" ("hole") algorithm; see [Dut87, She92, SMB94]. Like the non-decimated wavelet transform computed using nd.dwt, the à trous algorithm produces n wavelet coefficients at each multiresolution level for a signal with n sample values. The main difference is that the detail coefficients in the à trous algorithm are computed through simple differences between the smooth coefficients at different levels:

$$d_{j,k} = s_{j-1,k} - s_{j,k} . \qquad (11.1)$$

The detail coefficients produced by the nd.dwt function are computed using a dilated high-pass wavelet filter.

The à trous transform is computed with the function atrous. Compute the à trous transform for the glint noise signal as follows:

```
> glint.atrous <- atrous(glint, wavelet="s8", n.levels=5)
> glint.atrous
A Trous Wavelet Transform of glint
```

```
Non-Decimate Discrete Wavelet Transform of glint
Is an incomplete DWT
Wavelet: s8
Length of series: 512
Number of levels: 5
Boundary correction rule: periodic
Crystals: s5 d5 d4 d3 d2 d1
```

Compare the *à trous* transform with the transform produced by nd.dwt with:

```
> glint.nddwt <- nd.dwt(glint, wavelet="s8", n.levels=5)
> par(mfrow=c(1,2))
> plot(glint.atrous)
> plot(glint.nddwt)
```

The two transforms are displayed in figure 11.4.

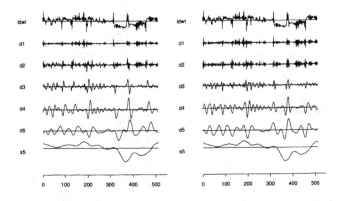

FIGURE 11.4. Comparison of non-decimated transforms for the radar glint noise signal. Left: the "*à trous*" wavelet transform. Right: the non-decimated wavelet transform.

11.2 Robust Smoother-Cleaner Wavelets

The robust smoother-cleaner wavelet transform, which was developed in [BDGM94], is a fast wavelet decomposition which is robust towards outliers. Smoother-cleaner wavelets behave like the classical L_2 wavelet transform for Gaussian signals, but prevent outliers and outlier patches from leaking into the wavelet coefficients at coarse levels.

To motivate the smoother-cleaner algorithm, use the WaveShrink estimator to smooth the glint noise signal. The WaveShrink estimator is discussed in chapter 6. We analyzed the glint noise in chapter 2, and the signal was shown in figure 4.1. Apply the `waveshrink` function to the glint noise signal:

```
> ts.plot(waveshrink(glint,shrink.rule="universal"))
```

The resulting estimated signal is shown in figure 11.5. The "true"

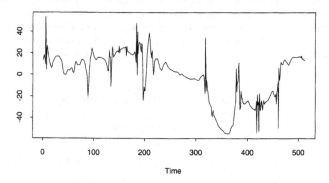

FIGURE 11.5. Time series plot for wavelet shrinkage of glint noise signal.

signal is a low-frequency oscillation around zero. The WaveShrink signal estimate is quite poor, as the glint spikes have leaked into the estimated signal.

The problem is that the glint spikes are highly non-Gaussian, and the noise model seriously deviates from the assumed Gaussian model (see chapter 6). The glint spikes correspond to outlier patches in the original signal. These are represented by large wavelet coefficients: see figure 4.2 for a plot of the DWT of the glint noise signal. Shrinking the coefficients does not remove the glint spikes.

The aim of the robust smoother-cleaner wavelet decomposition is to remove outlier patches from the wavelet decomposition. The noise in the robust wavelet decomposition is much closer to Gaussian, and the usual WaveShrink procedure can then be safely applied.

11.2.1 Smoother-Cleaner Algorithm

The basic idea of robust smoother-cleaner wavelets is simple: the smooth coefficients are preprocessed with a fast and robust smoother-cleaner. The procedure is illustrated in figure 11.6. As usual, we start with a set of wavelet coefficients s_0. Then, for each multiresolution

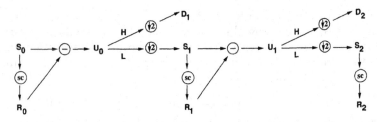

FIGURE 11.6. The robust smoother-cleaner algorithm produces a pyramid decomposition with an extra component: the robust residual r_j. For each multiresolution level, the low-pass coefficients s_j are first cleaned using a robust smoother-cleaner, denoted by **sc** in the figure. The residuals are saved in the r_j. The usual wavelet filters are then applied to the cleaned s_j to obtain s_{j+1} and d_{j+1}.

level, the signal is decomposed into three components:

1. A set of robust residuals r_{j-1}, given by

$$r_{j-1} = \delta_{\lambda_1,\lambda_2} (s_{j-1} - \widehat{s}_{j-1}) \qquad (11.2)$$

where $\delta_{\lambda_1,\lambda_2}$ is the semisoft shrinkage function (6.4) and $\widehat{s}_{j-1}$ is a robust smooth of s_{j-1} (e.g., running medians of 5). The thresholds (λ_1, λ_2) are chosen so that most of the robust residuals are zero.

2. The smooth wavelet coefficients s_j, which are obtained by applying the usual low-pass/decimation wavelet filter L to the cleaned smooth coefficients $u_{j-1} = s_{j-1} - r_{j-1}$.

3. The detail wavelet coefficients d_j obtained by applying the usual high-pass/decimation wavelet filter H to u_{j-1}.

The key properties of the robust smoother-cleaner wavelet transform are the following:

- In the Gaussian noise case, the robust smoother-cleaner wavelet transform produces essentially the same decomposition as the classical wavelet transform. By design, only a small number of robust residuals are detected, and these are small in magnitude.

- Outliers patches of length $2^\ell + 2$ are isolated to wavelet coefficients at levels $j \leq \ell$.

- Suppose the distribution of the noise F_ϵ has the form

$$F_\epsilon = (1 - \gamma)F + \gamma G \qquad (11.3)$$

where F is a Gaussian distribution, G is a "long-tailed" outlier producing distribution, and γ is the fraction of contamination. For outlier models of this type, it can be shown that WaveShrink, when applied in an appropriate way to the robust smoother-cleaner wavelet decomposition, can achieve a near optimality property similar to (6.1).

11.2.2 Computing the Robust Smoother-Cleaner DWT

Use the function `rob.dwt` to compute a robust smoother-cleaner DWT. For example, compute a robust DWT for the `glint` signal with:

```
> glint.robdwt <- rob.dwt(glint, wavelet="bs1.5")
> glint.robdwt
Robust Wavelet Transform for  glint
Wavelet: bs1.5
Length of series: 512
Number of levels: 6
Boundary correction rule: periodic
Crystals:   d1 d2 d3 d4 d5 d6 s6
Smoother Cleaner Residuals: r0 r1 r2 r3
```

The biorthogonal wavelet `bs1.5` is used here because it has short analysis filters; see section 11.2.3.

Plot the robust DWT and the corresponding multiresolution signal decomposition as follows:

```
> par(mfrow=c(1,2))
> plot(glint.robdwt)
> glint.rob <- decompose(glint.robdwt)
> plot(glint.rob)
```

Figure 11.7 shows the result. The robust residuals r_0, r_1, and r_2 all contain significant coefficients associated with the glint spikes. The corresponding robust residual signal components R_0, R_1, and R_2 capture significant signal variation, also associated with the glint spikes.

Now compare a series of robust multiresolution approximations to the classical L_2 multiresolution approximations. There is no special analogue to the `mra` function for the robust case, but you can

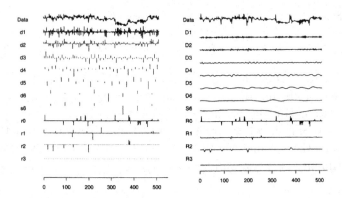

FIGURE 11.7. Robust smoother-cleaner wavelet decomposition (left) and the corresponding multiresolution signal decomposition (right) for the `glint` signal.

compute a series of robust multiresolution approximations from the decomposition object `glint.rob` as follows:

```
> rob.mra <- vector("list",5)
> Snms <- paste("S",1:4,sep="")
> Dnms <- paste("D",1:4,sep="")
> names(rob.mra) <- c("Data",Snms)
> rob.mra[["S4"]] <-
+    apply(glint.rob[c("D5", "D6", "S6")],1,sum)
> for(i in 4:2){
+    rob.mra[[Snms[i-1]]] <- rob.mra[[Snms[i]]] +
+        glint.rob[[Dnms[i]]]
+ }
> rob.mra[["Data"]] <- glint
```

Now plot the L_2 multiresolution approximations and the robust multiresolution approximations side-by-side to produce figure 11.8:

```
> par(mfrow=c(1,2))
> plot(mra(glint, n.level=4, wavelet="bs1.5"))
> stack.plot(rob.mra,same.scale=T)
```

The glint spikes are effectively removed from the robust approximations.

By using the robust smoother-cleaner DWT as a basis for wavelet shrinkage, we can obtain even better estimates of the signal. To see this, we first apply the `waveshrink` function to the robust DWT with the robust residuals removed:

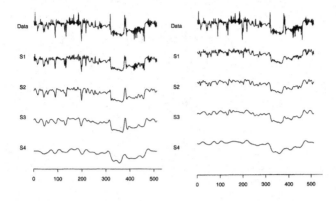

FIGURE 11.8. Multiresolution approximations for the `glint` signal. Left: L_2 approximation. Right: robust approximation.

```
> glint.clean1 <- as.dwt(glint.robdwt)
> glint.ws1 <- waveshrink(glint.clean1,
+   shrink.rule="universal")
```

The function `as.dwt` coerces the `glint.robdwt` as a DWT object, stripping off the robust residual terms. The estimate `glint.ws1` is obtained by shrinking the "clean" DWT coefficients. Now compare this robust estimate to the non-robust WaveShrink estimate:

```
> glint.ws0 <- waveshrink(glint, wavelet="bs1.5")
> stack.plot(list(Data=glint, DWT=glint.ws0,
+   RobDWT=glint.ws1), same.scale=T)
```

The result is shown in figure 11.9.   The robust estimate is not influenced by the glint spikes.

11.2.3   Selecting the Robust DWT

For the most part, the options for computing a robust smoother-cleaner DWT are the same as for the DWT. However, there are a few special considerations which you should take into account. These are discussed below.

Choice of Wavelet

The low-pass analysis filters should be short in order to avoid leakage of outlier patches to the smooth coefficients. In general, the longer the low-pass filter, the more an outlier patch tends to get smeared when going from fine to coarse levels. The smearing is undesirable

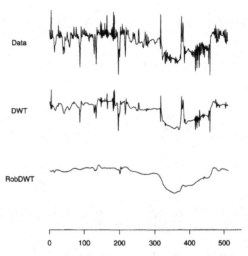

FIGURE 11.9. Comparison of robust and classical DWTs for the glint signal. Top: The original glint signal. Middle: the WaveShrink estimate based on the classical DWT. Bottom: the WaveShrink estimate based on the robust smoother-cleaner DWT.

since it then is difficult to isolate and identify the outlier patch (as in the L_2 case). On the other hand, it is desirable to have longer filters to ensure sufficient smoothness with the underlying basis functions.

The "b-spline" biorthogonal wavelets [Dau92] satisfy both requirements: short filters can be used for decomposition and longer filters for reconstruction. This is why the bs1.5 wavelet was used above for analysis of the glint noise signal.

Choice of Robust Filter

The smoother-cleaner step is based on a median filter. Median filters are computationally fast to compute, have a very high breakdown point, and are used extensively in the engineering community. (The *breakdown point* of an estimator is the largest fraction of data which may be replaced by arbitrarily large values without causing the Euclidean norm of the estimator to tend to infinity. The mean has breakdown point 0; the median approximately 1/2.)

The span of the median filter L should be as small as possible to provide minimal distortion of the underlying signal. However, L must be sufficiently big to prevent outlier patches from getting smeared in coarser scales. In theory, for a low-pass wavelet filter of length M, smearing is prevented by using median filters of length $L \geq 2M + 1$. In practice, using median filters of length 5 or 7 is usually sufficient to avoid smearing for most types of wavelets (the default is 5).

The robust smoother-cleaner step is implemented by the function `smoother.cleaner`.

Setting the Robust Residual Threshold

The thresholds (λ_1, λ_2) in (11.2) determine the number of non-zero robust residuals. Setting (λ_1, λ_2) too big results in leakage of outliers into the signal and setting (λ_1, λ_2) too small causes distortion of the signal. By default, a hard shrinkage function is used with a threshold of 2.5, so $(\lambda_1, \lambda_2) = (2.5, 2.5)$. This ensures that an average of 5% non-zero robust residuals remain after applying a threshold for Gaussian white noise. Refer to [BDGM94] for details.

11.3 Creating and Using Your Own Wavelets

New types of wavelets beyond those listed in chapter 2 are being developed. In this section, you will learn how to create you own wavelet functions and perform wavelet analyses using your own wavelet filters.

Warning: Analysis with your own wavelet filters is only partially supported by S+WAVELETS. You can plot wavelet functions and compute the 1-D DWT and wavelet packet transforms using your own wavelet filters. However, not all of the functions which apply to the standard wavelets of chapter 2 will work for your own wavelets.

11.3.1   Creating New Wavelets

To define your own wavelet function, you need to define a wavelet filter; see section 12.1. A wavelet filter consists of a vector of coefficients. For example, create and print the c6 low-pass wavelet filter:

```
> c6.lp <- wave.filter("c6")
> c6.lp
'c6'  Low Pass Wavelet Filter:
        1(-2)      1(-1)      1(0)       1(1)        1(2)
   -0.07273262 0.3378977 0.852572 0.3848648 -0.07273262
          1(3)
   -0.01565573
```

The c6 low-pass filter has 6 non-zero coefficients $\ell_{-2}, \ell_{-1}, \ldots, \ell_3$. Note that for coiflets, the index of the first filter coefficient is not zero. This index is important in defining the support of the wavelet function.

To create a wavelet function, instead of giving a wavelet name, you can supply a wavelet filter using the `filter` argument. If you supply a filter, then you should also supply the argument `filter.start`. For example, create the c6 coiflet as follows:

```
> c6wave <- wavelet(filter=c6.lp, mother=F,
+ filter.start=-2)
> c6wave
Father wavelet function
Wavelet name: user.defined
Level: 0
Shift: 0
```

To evaluate a wavelet function, use the `plot` function with the argument `plot=F`. Compare the values of the c6wave to those of the usual c6 wavelet:

```
> yval1 <- plot(wavelet("c6", mother=F), plot=F)$y
> yval2 <- plot(c6wave, plot=F)$y
> vecnorm(yval1 - yval2)/vecnorm(yval1)
[1] 9.956703e-16
```

A more demanding example is to create the biorthogonal wavelets given in table 8.5 of [Dau92] on page 284. These are biorthogonal wavelets which are close to coiflets.

For biorthogonal wavelets, you need to define both an analysis low-pass filter ℓ_n and a synthesis low-pass filter ℓ_n^*:

```
> bicoif.lp <-
+    c(-.000506524725, -.001266311813, .003837568681,
+       .015925480768,  -.039723557692, -.052305116758,
+       .286392513736, .575291895604, .286392513736,
+      -.052305116758, -.039723557692, .015925480768,
+       .003837568681, -.001266311813, -.000506524725)
> bicoif.lpdual <- c(.0125, -.03125, -.05, .28125, .575,
+       .28125, -.05, -.03125, .0125)
```

You don't need to create the high-pass filters since these are computed through the quadrature mirror filter relationship (12.3).

Create the father and mother wavelets associated with these filters as follows:

```
> bicoif.father <- wavelet(wavelet="bicoif",
+    filter=list(bicoif.lp, bicoif.lpdual),
+    mother=F, filter.start=c(-7, -4))
> bicoif.mother <- wavelet(wavelet="bicoif",
+    filter=list(bicoif.lp, bicoif.lpdual),
+    mother=T, filter.start=c(-7, -4))
```

Both the analysis and synthesis low-pass filters must be passed in. The `filter.start` argument gives the starting coefficient for the analysis and synthesis filters respectively. Plot these wavelets, shown in figure 11.10, as follows:

```
> par(mfrow=c(1,2))
> plot(bicoif.father)
> plot(bicoif.mother)
```

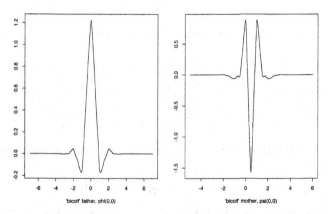

FIGURE 11.10. Biorthogonal wavelets close to coiflets given in table 8.5 of [Dau92] on page 284.

11.3.2 Analysis with Your Own Wavelet Functions

You can use the analysis functions `dwt`, `wp.table`, and `wpt` with your own wavelets. Take the DWT of a ramp signal using the `bicoif` wavelet:

```
> ramp <- make.signal("ramp", n=256)
> ramp.bicoif <- dwt(ramp, analysis.filter=bicoif.lp,
+      synthesis.filter=bicoif.lpdual,
+      boundary="reflection")
> plot(ramp.bicoif)
```

The plot of the transform is shown in figure 11.11. As with the **wavelet** function, you need to supply both the analysis and synthesis low-pass filters for biorthogonal wavelet analysis.

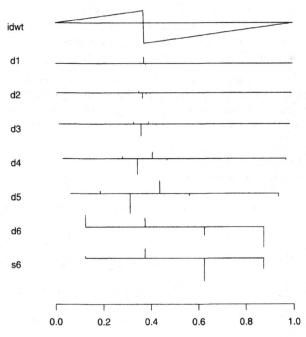

FIGURE 11.11. DWT of the ramp signal using the biorthogonal wavelets close to coiflets.

Now make sure that the transform is invertible:

```
> vecnorm((reconstruct(ramp.bicoif)-ramp))/vecnorm(ramp)
[1] 3.432636e-12
```

The numerical accuracy of the transform is roughly as good as the number of digits you provided for the filters.

Only two boundary rules can be used for wavelet analysis based on your own filters: **periodic** and **reflection**. The **periodic** boundary rule works only for sample sizes divisible by 2^J, where J is the maximum number of levels in the decomposition. The **reflection** boundary rule applies to all sample sizes, but gives perfect reconstruction only for biorthogonal symmetric filters. See chapter 14 for more information about boundary rules for wavelet analysis.

Note: For odd length biorthogonal wavelet filters having the **reflection** boundary rule, you need to specify the "correct" wavelet as the dual wavelet in order to get perfect reconstruction. In the above example, if the roles of `bicoef.lp` and `bicoef.lpdual` are switched, you can reconstruct the original signal only if you use the argument `dual=T`:

```
> ramp1 <- reconstruct(dwt(ramp,
+       analysis.filter=bicoif.lpdual,
+       synthesis.filter=bicoif.lp,
+       boundary="reflection"))
> ramp2 <- reconstruct(dwt(ramp, dual=T,
+       analysis.filter=bicoif.lpdual,
+       synthesis.filter=bicoif.lp,
+       boundary="reflection"))
> vecnorm(ramp1 - ramp)/vecnorm(ramp)
[1] 1.191008
> vecnorm(ramp2 - ramp)/vecnorm(ramp)
[1] 5.285793e-12
```

Hence, if you do not get perfect reconstruction with biorthogonal wavelets using the `reflection` boundary rule, try switching the `dual` argument.

12
Wavelet Algorithms and Filters

In section 2.6, you learned about the pyramid filtering algorithm for the discrete wavelet transform. In this chapter, you will learn more about wavelet filtering methods. Topics to be covered include:

- Creating wavelet filters with `wave.filter` and plotting their transfer functions with `transfer.plot` (section 12.1).

- Implementing the pyramid algorithm by means of the functions `convup` and `convdown` (section 12.2).

- Creating matrix wavelet operators using the `dwt.matrix` function (section 12.3).

- Properties of wavelet functions (section 12.4).

- The dilation equation and evaluating wavelet functions (section 12.5).

- Algorithms for the two-dimensional discrete wavelet transform (section 12.6).

12.1   Wavelet Filters

Both The discrete wavelet transform (DWT) and the inverse discrete wavelet transform (IDWT) use Mallat's [Mal89b] remarkably fast pyramid algorithms; see section 2.6. The forward algorithm involves use of *analysis* low-pass and high-pass filters L and H. The backward algorithm involves use of *synthesis* low-pass and high-pass filters L^* and H^*.

Only very special analysis filters L and H and synthesis filters L^* and H^* correspond to a discrete wavelet transform and its inverse [Dau92, Chu92a]. In this section, you will learn about these filters and some of their properties.

12.1.1   Orthogonal Wavelet Filters

The low-pass filter coefficients ℓ_n for an orthogonal discrete wavelet transform are determined by the father wavelets through the formula

$$\ell_n = \frac{1}{\sqrt{2}} \int \phi(t)\phi(2t - n)dt . \tag{12.1}$$

The high-pass filter coefficients h_n are determined by the inner product of the father and mother wavelets through the formula

$$h_n = \frac{1}{\sqrt{2}} \int \psi(t)\phi(2t - n)dt . \tag{12.2}$$

The low- and high-pass filter coefficients are directly related by the *quadrature mirror filter* relationship:

$$h_n = (-1)^n \ell_{1-n} . \tag{12.3}$$

The coefficients ℓ_n and h_n are called the *impulse responses* of the low- and high-pass filters respectively.

Note: The wavelet filter coefficients have been scaled so that $\sum_n \ell_n^2 = \sum_n h_n^2 = 1$. In the book by Daubechies [Dau92], the filter coefficients are scaled so that $\sum_n \ell_n^2 = \sum_n h_n^2 = 2$.

You can obtain the low-pass and high-pass filters with the function `wave.filter`:

```
> s8.lp <- wave.filter("s8")
> s8.hp <- wave.filter("s8",high.pass=T)
> s8.lp
```

```
's8'  Low Pass Wavelet Filter:
       1(0)          1(1)          1(2)         1(3)          1(4)
  0.0322231 -0.01260397 -0.09921954 0.2978578 0.8037388
       1(5)          1(6)          1(7)
  0.4976187 -0.02963553 -0.07576571
> s8.hp
's8'  High Pass Wavelet Filter:
       h(-6)         h(-5)         h(-4)        h(-3)         h(-2)
  0.07576571 -0.02963553 -0.4976187 0.8037388 -0.2978578
       h(-1)         h(0)          h(1)
 -0.09921954 0.01260397 0.0322231
```

Plot the coefficients as follows to obtain figure 12.1:

```
> par(mfrow=c(1,2))
> plot(s8.lp)
> plot(s8.hp)
```

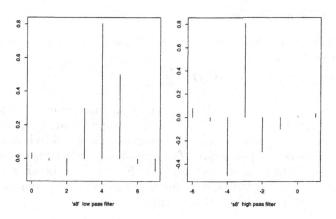

FIGURE 12.1. Coefficients of the low-pass and high-pass filters for the **s8** wavelet.

For the orthogonal wavelets in S+WAVELETS, the synthesis filters L^* and H^* are the same as the analysis filters L and H:

$$\ell_n^* = \ell_n$$
$$h_n^* = h_n$$

12.1.2   Convolution Formulas

The forward pyramid algorithm used to compute the discrete wavelet transform, described in section 2.6, is implemented by a series of filtering and down-sampling operators. The jth step in the forward pyramid algorithm computes the level j smooth and detail coefficients $s_{j,k}$ and $d_{j,k}$ from the level $j-1$ smooth coefficients $s_{j-1,k}$ as follows:

$$s_{j,k} = \sum_n \ell_{2k-n} s_{j-1,n} \qquad (12.4)$$

$$d_{j,k} = \sum_n h_{2k-n} s_{j-1,n} \qquad (12.5)$$

The jth step in the backwards pyramid algorithm computes the level $j-1$ smooth coefficients from the level j coefficients $s_{j,k}$ and $d_{j,k}$ as follows:

$$s_{j-1,n} = \sum_k (\ell_{n-2k} s_{j,k} + h_{n-2k} d_{j,k}) \ . \qquad (12.6)$$

12.1.3 Transfer Functions

The *transfer function* $G(\omega)$ for a filter g with impulse response g_n is given by

$$G(\omega) = \sum_n g_n e^{-i\omega n} \ .$$

The transfer function indicates the "frequency domain" effects of a filter applied to a signal. For a given input sinusoid signal with frequency ω, the transfer function describes the amplitude and phase of the filtered output signal. The amplitude for a transfer function of a wavelet filter is plotted using the function **transfer.plot**. For example, compare the amplitudes of the transfer functions for a pair of low- and high-pass s8 filters with:

```
> par(mfrow=c(2,1))
> transfer.plot(s8.lp)
> transfer.plot(s8.hp)
```

The plot is shown in figure 12.2.

The phase of the transfer function is obtained by setting the argument **phase=T** in **transfer.plot**. The phases of the low-pass s8 wavelet filter and d8 wavelet filter are compared in figure 12.3 using:

```
> par(mfrow=c(2,1))
> transfer.plot(s8.lp,phase=T)
> transfer.plot(wave.filter("d8"),phase=T)
```

The phase plot shows that the s8 wavelet filter, which is nearly symmetric, has nearly linear phase. By contrast, the phase of the d8

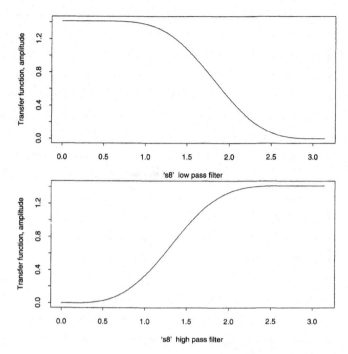

FIGURE 12.2. Amplitude of the transfer functions of the low and high pass filters for s8 wavelet.

wavelet filter, which is highly asymmetric, is non-linear. The discontinuities in phase plots are a "wrap-around" effect. This is because the arctan function, which computes the phase, computes the true angle modulus 2π.

Next we apply `transfer.plot` to a DWT object created using the s8 filter:

```
> glint.dwt <- dwt(glint, wavelet="s8")
> transfer.plot(glint.dwt)
```

The result is shown in figure 12.4. The transfer function in the top plot (labeled H) corresponds to the high pass filter which produces the detail coefficients d_1, and the transfer function in the second plot (labeled HL) corresponds to the bandpass filter which produces the detail coefficients d_2, and so on.

The frequency bandwidth of the filters gets smaller for coarser scales. Wavelet filters have the *constant-Q* property, which states that the ratio of the filter bandwidth to the center frequency is constant. This corresponds to better scale localization at coarse scales

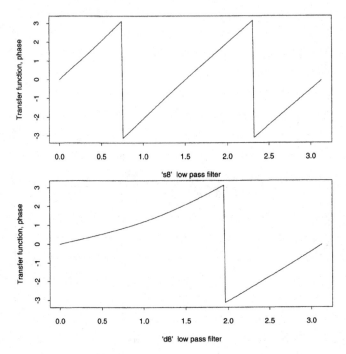

FIGURE 12.3. Phase of the transfer functions of the low pass filters for s8 wavelet and d8 wavelet.

and better time localization at fine scales (see section 2.4).

Plot the transfer functions using a haar filter as follows:

```
> glint.haar <- dwt(glint, wavelet="haar")
> transfer.plot(glint.haar)
```

This is shown in figure 12.5. The haar bandpass filters have very significant ripples or sidelobes. Also, the shape of the haar filters is less sharp than that of the s8 filters. These are both reasons for preferring the s8 wavelet (and other smooth wavelets) over the haar wavelet.

12.1.4 Biorthogonal Wavelet Filters

The analysis filters L and H for biorthogonal wavelets can be obtained just as for orthogonal wavelets using (12.1) and (12.2). However, unlike orthogonal wavelets, the synthesis filters L^* and H^* for biorthogonal wavelets are not the same as the analysis filters. The impulse responses of the synthesis filters L^* and H^* are derived from

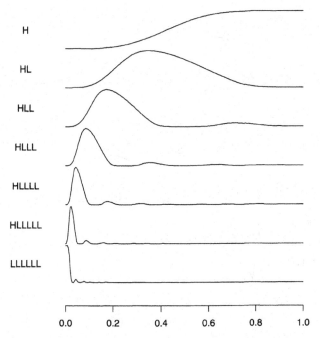

FIGURE 12.4. Transfer functions of the **s8** bandpass DWT filters.

the dual wavelet functions:

$$\ell_n^* = \frac{1}{\sqrt{2}} \int \widetilde{\phi}(t)\widetilde{\phi}(2t - n)dt \qquad (12.7)$$

$$h_n^* = \frac{1}{\sqrt{2}} \int \widetilde{\psi}(t)\widetilde{\phi}(2t - n)dt \ . \qquad (12.8)$$

The analogue to the quadrature mirror filter relationship (12.3) for biorthogonal wavelets is

$$h_n = (-1)^n \ell_{1-n}^*$$
$$h_n^* = (-1)^n \ell_{1-n} \ .$$

Use the **wave.filter** function to create biorthogonal wavelet filters. The synthesis filters are obtained with the argument **dual=T**. For example, we plot the analysis and synthesis biorthogonal filters for the **bs3.5** wavelet as follows:

```
> par(mfrow=c(2,2))
> xlim <- c(-5,6)
> ylim <- c(-1,1)
```

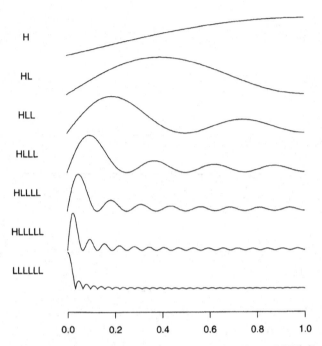

FIGURE 12.5. Transfer functions of the **haar** bandpass DWT filters.

```
> plot(wave.filter("bs3.5"), xlim=xlim, ylim=ylim)
> abline(h=0, lty=2)
> plot(wave.filter("bs3.5", high=T), xlim=xlim,
+      ylim=ylim)
> abline(h=0, lty=2)
> plot(wave.filter("bs3.5", dual=T), xlim=xlim,
+      ylim=ylim)
> abline(h=0, lty=2)
> plot(wave.filter("bs3.5", high=T, dual=T),
+      xlim=xlim, ylim=ylim)
> abline(h=0, lty=2)
```

The plot is shown in figure 12.6. An important feature of biorthogonal wavelet filters is their symmetry.

Plot the transfer function for the **bs3.5** low-pass analysis filter:

```
> par(mfrow=c(1,2))
> transfer.plot(wave.filter("bs3.5"))
> transfer.plot(wave.filter("bs3.5"),phase=T)
```

which results in figure 12.7. Since the **bs3.5** filter is symmetric, it has exactly linear phase.

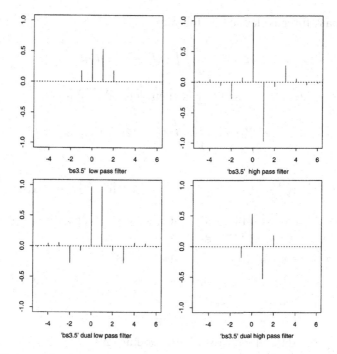

FIGURE 12.6. The analysis and synthesis biorthogonal filters for the **bs3.5** wavelet.

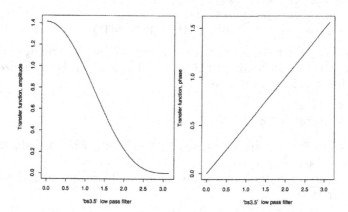

FIGURE 12.7. Transfer function for the analysis biorthogonal **bs3.5** filter.

Note: For biorthogonal wavelets, you can reverse the roles of the wavelets ϕ, ψ and the dual wavelets $\tilde{\phi}$, $\tilde{\psi}$. In this case, the analysis low-pass filter is defined as follows:

$$\ell_n = \frac{1}{\sqrt{2}} \int \widetilde{\phi}(t)\widetilde{\phi}(2t - n)dt \ . \tag{12.9}$$

The synthesis low-pass filter is then defined as follows:

$$\ell_n^* = \frac{1}{\sqrt{2}} \int \phi(t)\phi(2t - n)dt \ . \tag{12.10}$$

12.2 Implementing the Pyramid Algorithm

In this section, we will implement the pyramid algorithms "by hand." First, create a `doppler` signal vector:

```
> x <- make.signal("doppler", n=64)
```

Now create the low-pass and high-pass "s8" wavelet filters:

```
> L <- wave.filter("s8")
> H <- wave.filter("s8",high=T)
```

Set up the data structures for the coefficients:

```
> s <- vector("list", 4)
> names(s) <- c("s0", "s1", "s2", "s3")
> d <- vector("list", 4)
> names(d) <- c("d0", "d1", "d2", "d3")
```

The object s is for the smooth coefficients s_0, s_1, s_2, and s_3 and the object d is for the detail coefficients d_0, d_1, d_2, and d_3. Initialize the level 0 coefficients as follows:

```
> s[["s0"]] <- x
> d[["d0"]] <- x*0
```

The s_0 coefficients are initialized with the data while the d_0 coefficients are set to zero.

Now run three iterations of the pyramid algorithm using the function `convdown`:

```
> for(j in 1:3){
+    d[[j+1]] <- convdown(s[[j]], filter=H,
+        boundary="periodic")
+    s[[j+1]] <- convdown(s[[j]], filter=L,
+        boundary="periodic")
+ }
```

The convdown function implements the combined step of filtering and down-sampling by 2. A plot of the coefficients with the function stack.plot is shown in figure 12.8:

```
> par(mfrow=c(1,2))
> stack.plot(s, type="h", zerocenter=T)
> stack.plot(d, type="h", zerocenter=T)
```

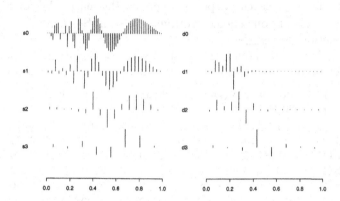

FIGURE 12.8. Illustration of the pyramid algorithm.

The convup function implements the combined filtering and up-sampling by 2. Recall that for orthogonal wavelets, the synthesis filters L^* and H^* are the same as the analysis filters. Hence, the backwards pyramid-reconstruction algorithm is given by

```
> y <- s[[4]]
> j <- 4
> for(j in 4:2){
+     y <- convup(y, filter=L, boundary="periodic") +
+          convup(d[[j]], filter=H, boundary="periodic")
+ }
> vecnorm(x - y)/vecnorm(x)
[1] 8.390169e-13
```

The relative L^2 error of our reconstruction is less than 10^{-12}.

12.3   The DWT as a Linear Transform

Mallat's pyramid algorithm is a fast way to compute the discrete wavelet transform (DWT). As mentioned in section 2.2, DWT is mathematically equivalent to the matrix multiplication

$$w = Wf$$

where W is an orthogonal matrix (when the periodic boundary rule is used). The pyramid algorithm does not explicitly perform this matrix calculation. However, you may find it useful, for either instructional or research purposes, to create the matrix W for a given wavelet basis of your choice. You may do so in S+WAVELETS using the function dwt.matrix.

For example, we create the wavelet transform matrix for the haar wavelet with n.level=3 levels and a data vector of length 8 with:

```
> W <- dwt.matrix(wavelet="haar", n=8, n.level=3)
> round(W,2)
       [,1]  [,2]  [,3]  [,4]  [,5]  [,6]  [,7] [,8]
1(1)   0.35  0.35  0.35  0.35  0.35  0.35  0.35 0.35
h(1)  -0.35 -0.35 -0.35 -0.35  0.35  0.35  0.35 0.35
h(1)  -0.50 -0.50  0.50  0.50  0.00  0.00  0.00 0.00
h(2)   0.00  0.00  0.00  0.00 -0.50 -0.50  0.50 0.50
h(1)  -0.71  0.71  0.00  0.00  0.00  0.00  0.00 0.00
h(2)   0.00  0.00 -0.71  0.71  0.00  0.00  0.00 0.00
h(3)   0.00  0.00  0.00  0.00 -0.71  0.71  0.00 0.00
h(4)   0.00  0.00  0.00  0.00  0.00  0.00 -0.71 0.71
```

The first row corresponds to the wavelet smooth coefficient $s_{3,1}$ at level 3 (scale $2^3 = 8$). The second row corresponds to the wavelet detail coefficient $d_{3,1}$ at level 3. The third and fourth rows correspond to detail coefficients $d_{2,1}$ and $d_{2,2}$ at level 2. The last four rows correspond to detail coefficients $d_{1,1}$, $d_{1,2}$, $d_{1,3}$, and $d_{1,4}$ at level 1.

Since W is an orthogonal matrix, the inverse DWT can be obtained by taking the transpose matrix $W' = W^{-1}$. You compute $W'W$ as follows, using the matrix multiplication operator %*%:

```
> t(W) %*% W
                [,1]             [,2]             [,3]
[1,]   1.000000e+00   -1.110223e-16     0.000000e+00
[2,]  -1.110223e-16    1.000000e+00     0.000000e+00
[3,]   0.000000e+00    0.000000e+00     1.000000e+00
[4,]   0.000000e+00    0.000000e+00    -1.110223e-16
[5,]   0.000000e+00    0.000000e+00     0.000000e+00
[6,]   0.000000e+00    0.000000e+00     0.000000e+00
[7,]   0.000000e+00    0.000000e+00     0.000000e+00
[8,]   0.000000e+00    0.000000e+00     0.000000e+00
```

```
                  [,4]             [,5]             [,6]
[1,]     0.000000e+00    0.000000e+00    0.000000e+00
[2,]     0.000000e+00    0.000000e+00    0.000000e+00
[3,]    -1.110223e-16    0.000000e+00    0.000000e+00
[4,]     1.000000e+00    0.000000e+00    0.000000e+00
[5,]     0.000000e+00    1.000000e+00   -1.110223e-16
[6,]     0.000000e+00   -1.110223e-16    1.000000e+00
[7,]     0.000000e+00    0.000000e+00    0.000000e+00
[8,]     0.000000e+00    0.000000e+00    0.000000e+00
                  [,7]             [,8]
[1,]     0.000000e+00    0.000000e+00
[2,]     0.000000e+00    0.000000e+00
[3,]     0.000000e+00    0.000000e+00
[4,]     0.000000e+00    0.000000e+00
[5,]     0.000000e+00    0.000000e+00
[6,]     0.000000e+00    0.000000e+00
[7,]     1.000000e+00   -1.110223e-16
[8,]    -1.110223e-16    1.000000e+00
```

Compare the direct use of the matrix transformation $\mathbf{w} = \mathbf{Wf}$ with use of the dwt function:

```
> f <- 1:8
> W %*% f
              [,1]
l(1)   12.7279221
h(1)    5.6568542
h(1)    2.0000000
h(2)    2.0000000
h(1)    0.7071068
h(2)    0.7071068
h(3)    0.7071068
h(4)    0.7071068
> as.matrix(dwt(f, wavelet="haar", boundary="periodic"))
              [,1]
s3(1)  12.7279221
d3(1)   5.6568542
d2(1)   2.0000000
d2(2)   2.0000000
d1(1)   0.7071068
d1(2)   0.7071068
d1(3)   0.7071068
d1(4)   0.7071068
```

The S-PLUS function as.matrix coerces a vector of length n into an $n \times 1$ matrix.

Warning: For n.level > 1, the function dwt.matrix produces transform matrices using the **periodic** boundary correction rule. To appropriately compare the use of the transform matrix W to the use of the dwt function with n.level > 1, you need to use the dwt optional argument boundary="periodic". See section 14.4.1 for matrix operators corresponding to boundary corrections other than **periodic**.

12.4   Properties of Wavelet Functions

Selecting a wavelet requires a tradeoff between different properties such as smoothness, spatial localization, frequency localization, the ability to represent local polynomial functions, orthogonality, and symmetry. These properties are discussed below:

- *Smoothness*: The smoothness of the wavelet approximation is one of the properties that distinguishes modern wavelet analysis. For many applications, the wavelet function must be sufficiently smooth to efficiently represent the characteristics of the underlying signal. The lack of smoothness is one of the main disadvantages of the **haar** wavelet.

  One measure of smoothness for a wavelet is given by the number of derivatives which exist for that wavelet. The **haar** wavelet is discontinuous, and hence is not differentiable. The **d4** wavelet is continuous but also is not differentiable. The **d12** wavelet, however, is twice differentiable.

- *Temporal/Spatial Localization*: A central feature of wavelet analysis is the ability to localize features in time and space. The support width of a wavelet is closely related to its ability to localize features in time and space. Very compact wavelets, such as the **haar** wavelet, are very well localized in time and space. Support width is generally inversely related to the smoothness; the smoothest wavelets have the widest support width.

- *Vanishing Moments*: A wavelet with a higher number of "vanishing moments" can better represent higher degree polynomial signals. The number of vanishing moments are also closely related to the smoothness of a wavelet. A mother wavelet ψ with

M vanishing moments satisfies

$$\int t^m \psi(t)dt = 0 \qquad m = 0, 1, \ldots, M - 1 . \qquad (12.11)$$

The *coiflet* has the unusual property of also having vanishing moments for the father wavelet ϕ:

$$\int t^m \phi(t)dt = 0 \qquad m = 1, \ldots, M - 1 .$$

The zero moment for the father coiflet is always one.

- *Frequency Localization*: Wavelets localize features not only in time and space, but also in frequency. The **haar** wavelet has very poor frequency resolution. In general, smoother wavelets have better frequency localization properties.

- *Symmetry*: With the exception of the **haar** wavelet, the orthogonal wavelets which have compact support are not symmetric; the *daublets* are highly asymmetric and the *symmlets* and *coiflets* are nearly symmetric. All of the biorthogonal wavelets are either symmetric or anti-symmetric. Symmetric wavelets have the advantage of avoiding any phase shifts; the wavelet coefficients do not "drift" relative to the original signal.

- *Orthogonality*: The orthogonality of the wavelet transform is a central feature for some applications. The biorthogonal wavelets lack the orthogonality property, although the *v-spline* biorthogonal wavelets are nearly orthogonal.

To get information about these properties of a wavelet, use the **summary** function. For example, look at the **c12** mother wavelet with:

```
> summary(wavelet("c12"))

c12 Mother Wavelet Function: Psi0.0
support:  [-5, 6]
level shift scale location
   0    0    1       0
length vanish.moments n.derivatives orthogonality
 "12"    "3"             "1"            "TRUE"
symmetry/anti-symmetry
"FALSE"
```

The frequency properties of a wavelet can be assessed by looking at its Fourier transform, defined by

$$\widehat{\psi}(\omega) = \frac{1}{\sqrt{2\pi}} \int_{-\infty}^{\infty} \psi(x)e^{-i\omega x}dx \ . \tag{12.12}$$

The Fourier transform of a wavelet is plotted in S+WAVELETS with the function `fourier.transform.plot`. For example, compare the Fourier transform of the `haar` and `s8` wavelets using:

```
> par(mfrow=c(1,2))
> fourier.transform.plot(wavelet("haar", mother=F))
> fourier.transform.plot(wavelet("s8", mother=F))
```

The resulting plot is shown in figure 12.9. The `haar` wavelet has much poorer frequency localization than the `s8` symmlet as evidenced by the less sharp drop-off at the central lobe and the larger sidelobes of the Fourier transform. In fact, of all the wavelets in S+WAVELETS, the `haar` wavelet has the poorest frequency localization.

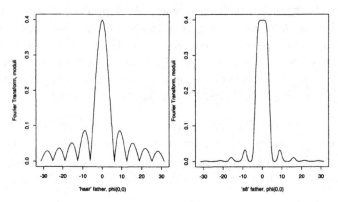

FIGURE 12.9. Comparing wavelet Fourier transforms. Left: the `haar` wavelet. Right: the `s8` wavelet.

Turn to appendices A and B for a plot of all orthogonal wavelets and biorthogonal wavelets in S+WAVELETS. Included in these appendices are tables summarizing the key properties of these wavelets.

12.5   The Dilation Equation

Except for a few special cases, wavelet functions have no analytic form. To evaluate a wavelet function, use the *dilation equation*, which lies at the heart of wavelet analysis. For a father wavelet $\phi(x)$, the dilation equation is defined by

$$\phi(x) = \sqrt{2}\sum_k \ell_k\phi(2x - k) . \tag{12.13}$$

You can obtain the mother wavelet $\psi(x)$ from the father wavelet by the relationship

$$\psi(x) = \sqrt{2}\sum_k h_k\phi(2x - k) . \tag{12.14}$$

The coefficients ℓ_k and h_k are the low-pass and high-pass filter coefficients defined in (12.1) and (12.2) respectively.

The dilation equation forms the basis for most wavelet calculations, including the forward and backward pyramid algorithms. To learn more about the dilation equation and its relationship to the pyramid algorithms, refer to a wavelet tutorial article [Str89, JS94a] or the books by Daubechies [Dau92] and Chui [Chu92a].

12.5.1   Evaluating a Wavelet Function at the Dyadics

One way to evaluate a wavelet function is to start with the values $\ldots, \phi(-1), \phi(0), \phi(1),\ldots$ of the wavelet at the integers. Then use the dilation equation (12.13) to refine $\phi(x)$ to half integers $x = 2^{-1}k$ for k any integer. Iterate (12.13) j times to evaluate $\phi(x)$ on a grid with a sampling interval of 2^{-j}. (Such a grid is called a *dyadic grid* because every point on the grid is a rational number having a power of 2 as its denominator.) This method for evaluating a wavelet function is equivalent to "construction 3" of [Str89].

To illustrate "turning the crank" of the dilation equation, we evaluate a wavelet function on successively finer dyadic grids, starting at the integers ($j = 0$). The argument J=j to the `plot` function applied to a wavelet object evaluates the wavelet on a dyadic grid with sampling interval 2^{-j}. Use the `plot` function to evaluate the d4 wavelet on successively finer dyadic intervals, as shown in figure 12.10:

```
> par(mfrow=c(4,2))
> d4 <- wavelet("d4", mother=F)
> for(j in 0:7){
+    plot(d4, J=j, type="p",
+         xlab=paste("Scale 2^-",j,sep=""))
+    abline(h=0, lty=2)
+ }
```

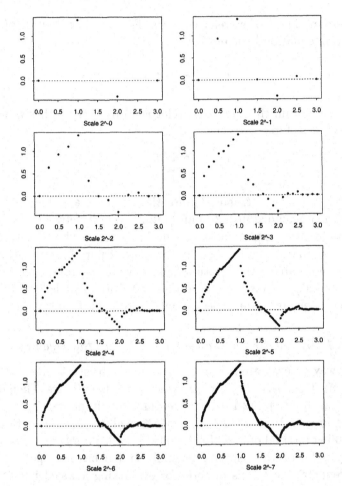

FIGURE 12.10. The dilation equation is used to compute the dyadic refinement of the d4 wavelet.

We start off at sampling interval $j = 2^0$ with the d4 wavelet evaluated at the integers. One iteration of the dilation equation yields the d4 wavelet refined to the half integers (sampling interval $j = 2^{-1}$). The wavelet is refined to the next finer dyadic sampling interval with each iteration of the dilation equation.

12.5.2   Evaluating a Wavelet Function at the Integers

To initialize the dyadic refined procedure of the previous section, we need to evaluate the wavelet function at the integers. This is done

by solving an eigenvector equation based on the dilation equation
(12.13) [Str89]. For the **d4** wavelet, the eigenvector equation is

$$\begin{pmatrix} \phi(1) \\ \phi(2) \end{pmatrix} = \sqrt{2} \begin{pmatrix} \ell_1 & \ell_0 \\ \ell_3 & \ell_2 \end{pmatrix} \begin{pmatrix} \phi(1) \\ \phi(2) \end{pmatrix}$$

Use the function `phi.at.integers` to solve this eigenvector equation.

Warning: The `bs2.2`, `bs3.1`, `bs3.3`, and `vs3` dual wavelet functions are infinite at all dyadic rationals! Hence, the above eigenequation approach does not work. The plots for these wavelets only approximate the shape and do not give the true values. See [RBG94] for details.

12.6   Algorithms for the 2-D DWT

The two-dimensional pyramid algorithm used to compute the 2-D discrete wavelet transform is a straightforward generalization of the 1-D DWT pyramid algorithm. Figure 12.11 displays one stage of the 2-D pyramid algorithm.

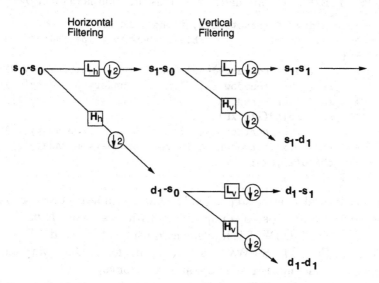

FIGURE 12.11. 2-D DWT pyramid algorithm.

The input to the 2-D pyramid algorithm consists of the values of the discrete image s0-s0 $= F$ where

$$F = F_{m,n} \qquad m = 1, 2, \ldots, M, \quad n = 1, 2, \ldots, N .$$

The algorithm consists of filtering and down-sampling *horizontally* using low- and high-pass filters L_h and H_h (the subscript h stands for "horizontal"). Horizontal filtering applies L_h and H_h to each row in the image and produces the coefficient matrices s1-s0 and d1-s0. Vertical filtering and down-sampling follows, using low- and high-pass filters L_v and H_v. The vertical filtering is applied to each of the coefficient matrices s1-s0 and d1-s0 to produce the set of 2-D wavelet coefficient matrices:

$$
\begin{aligned}
(\text{s1} - \text{s1})_{m,n} &= s_{1,m,n} \\
(\text{s1} - \text{d1})_{m,n} &= d^h_{1,m,n} \qquad m = 1, 2, \cdots, M/2 \\
(\text{d1} - \text{s1})_{m,n} &= d^v_{1,m,n} \qquad n = 1, 2, \cdots, N/2 \\
(\text{d1} - \text{d1})_{m,n} &= d^d_{1,m,n}
\end{aligned}
$$

The 2-D pyramid algorithm then iterates on the s1-s1 coefficient matrix.

One stage in the 2-D pyramid algorithm is implemented with the S+WAVELETS function quad.down. This function is defined by:

```
> quad.down <- function(x, fl.hor, fh.hor, fl.ver,
+                       fh.ver, boundary="periodic")
+ {
+   if(length(boundary)==1) boundary <- rep(boundary,2)
+   xl <- hor.convdown(x, fl.hor, boundary=boundary[1])
+   xh <- hor.convdown(x, fh.hor, boundary=boundary[1])
+   x <- rbind(xl, xh)
+   xl <- ver.convdown(x, fl.ver, boundary=boundary[2])
+   xh <- ver.convdown(x, fh.ver, boundary=boundary[2])
+   cbind(xl, xh)
+ }
```

The horizontal filtering step is implemented with hor.convdown and the vertical filtering step is implemented with ver.convdown.

The inverse 2-D DWT is implemented using a backward 2-D pyramid algorithm. The S+WAVELETS functions for the backward algorithm are quad.up, hor.convup, and ver.convup.

A one-level 2-D DWT decomposition is equivalent to the matrix multiplication

$$\mathbf{w} = \mathbf{W}_h F \mathbf{W}'_v$$

where $\mathbf{W}_h$ and $\mathbf{W}_v$ are orthogonal wavelet matrix operators. You can produce $\mathbf{W}_h$ and $\mathbf{W}_v$ using the dwt.matrix function (see section 12.3). For example, compute the one-level 2-D DWT using matrix operators for a simple 8×8 matrix as follows:

```
> W <- dwt.matrix(wavelet="haar", n=8, n.level=1)
> mat <- matrix(1:64,8,8)
> round(W %*% mat %*% t(W), 2)
```

	l(1)	l(2)	l(3)	l(4)	h(1)	h(2)	h(3)	h(4)
l(1)	11	43	75	107	8	8	8	8
l(2)	15	47	79	111	8	8	8	8
l(3)	19	51	83	115	8	8	8	8
l(4)	23	55	87	119	8	8	8	8
h(1)	1	1	1	1	0	0	0	0
h(2)	1	1	1	1	0	0	0	0
h(3)	1	1	1	1	0	0	0	0
h(4)	1	1	1	1	0	0	0	0

Compare this to the matrix produced by the dwt.2d function:

```
> round(as.matrix(dwt.2d(mat, wavelet="haar",
+ n.levels=1)),2)
```

	s1(1)	s1(2)	s1(3)	s1(4)	d1(1)	d1(2)	d1(3)	d1(4)
s1(1)	11	43	75	107	8	8	8	8
s1(2)	15	47	79	111	8	8	8	8
s1(3)	19	51	83	115	8	8	8	8
s1(4)	23	55	87	119	8	8	8	8
d1(1)	1	1	1	1	0	0	0	0
d1(2)	1	1	1	1	0	0	0	0
d1(3)	1	1	1	1	0	0	0	0
d1(4)	1	1	1	1	0	0	0	0

To compute a second level of the 2-D DWT using matrix operators, it is necessary to subset the s1-s1 matrix and then apply new $n/2 \times n/2$ wavelet operators.

13

Wavelet Packet and Cosine Packet Algorithms

Like the discrete wavelet transform (DWT), wavelet packet and cosine packet transforms are computed through a collection of efficient algorithms. The pyramid algorithm for the DWT extends in
a straightforward manner to wavelet packet analysis. Cosine packet
transforms are based on the discrete cosine transform (DCT). This
chapter, in which you will learn more about these algorithms, covers
the following topics:

- Wavelet packet filters, transfer functions, and algorithms for
  wavelet packets (section 13.1).

- Algorithms for cosine packet analysis (section 13.2).

- Extending 1-D algorithms to obtain algorithms for 2-D wavelet
  packets and cosine packets (section 13.3).

For a more comprehensive discussion of algorithms for wavelet
packet and cosine packet analysis, refer to Wickerhauser's book,
Adapted Wavelet Analysis—from theory to software [Wic94a]. For
details concerning algorithms for the DCT, refer to the book by Rao
and Yip, *Discrete Cosine Transform: Algorithms, Advantages, Applications* [RY90].

13.1   Wavelet Packet Filters and Algorithms

Section 2.6 and chapter 12 presented the fast DWT pyramid algorithm, wavelet filters and transfer functions, and the dilation equation. These algorithms can be extended in a straightforward manner for wavelet packet analysis. This section gives a brief overview of the computational aspects of wavelet packet analysis.

13.1.1   The Splitting Algorithm

A wavelet packet table computed by the wp.table function uses a fast splitting algorithm which is an adaptation of the pyramid algorithm for the DWT. As with the pyramid algorithm, the splitting algorithm involves use of low-pass and high-pass filters, along with a *down-sampling* (decimation) operator. The splitting algorithm differs from the pyramid algorithm in that:

- Low- and high-pass filters are applied to the detail coefficients in addition to the smooth coefficients at each stage in the algorithm.

- All coefficients are retained, including those at intermediate filtering stages.

Table 13.1 displays the filtering sequences used to obtain a wavelet packet table with 3 resolution levels. The level 0 coefficients are initialized with the signal f_k:

$$w_{0,0,k} = f_k \qquad k = 1, 2, \ldots, n.$$

The level 1 coefficients are obtained from $\mathbf{w}_{0,0}$ just as in the pyramid algorithm; a low-pass filter L is applied followed by down-sampling by 2 to obtain $\mathbf{w}_{1,0}$. A high-pass filter H, and down-sampling by 2 are applied to obtain $\mathbf{w}_{1,1}$.

As in the pyramid algorithm, the level 2 coefficients $\mathbf{w}_{2,0}$ and $\mathbf{w}_{2,1}$ are obtained by applying low- and high-pass filters to $\mathbf{w}_{1,0}$. The

0	Data							
1	L				H			
2	LL		HL		HH		LH	
3	LLL	HLL	HHL	LHL	LHH	HHH	HLH	LLH

TABLE 13.1. The filters applied to produce a wavelet packet table with 3 resolution levels.

splitting algorithm differs from the pyramid algorithm in computing the level 2 coefficients $\mathbf{w}_{2,2}$ and $\mathbf{w}_{2,3}$. These are obtained by applying high- and low-pass filters to "detail" coefficients $\mathbf{w}_{1,1}$ (in the pyramid algorithm, no further filter is applied to these coefficients). Note the reversal of high- and low-pass filters: the high-pass filter is used to obtain $\mathbf{w}_{2,2}$ while the low-pass filter is used to obtain $\mathbf{w}_{2,3}$

The splitting proceeds down the table, filtering the "parent" coefficients $\mathbf{w}_{j,b}$ with low- and high-pass filters to obtain the coefficients $\mathbf{w}_{j+1,2b}$ and $\mathbf{w}_{j+1,2b+1}$. The choice of low-pass or high-pass filter is determined by the block number b of the parent: if b is even, then a low-pass filter is used to obtain $\mathbf{w}_{j+1,2b}$ and a high-pass filter is used to obtain $\mathbf{w}_{j+1,2b+1}$; otherwise, the low-pass and high-pass filters are reversed. This filtering sequence is known as *sequency* order, and ensures that the blocks have increasing frequency (oscillation) going from the left to the right in the table. See the book by Wickerhauser [Wic94a] for details.

The algorithm to compute the wavelet packet table has complexity $O(n \log n)$.

13.1.2   Wavelet Packet Filters

A wavelet packet filter can be obtained in S+WAVELETS with the `wave.filter` function using the argument `filter.seq`, which specifies the series of filters to apply. For example, create and plot the wavelet packet filter as follows:

```
> wp.filt <- wave.filter("s8", filter.seq="LLH")
> plot(wp.filt)
```

This filter corresponds to the wavelet packet crystal $\mathbf{w}_{3,7}$. The impulse response of the filter is shown in figure 13.1.

Plot the amplitude of the transfer function using the function `transfer.plot`, to create figure 13.2:

```
> transfer.plot(wp.filt)
```

You can also compute the transfer functions corresponding to the filters used to compute a wavelet packet transform. Compute the transfer functions for the level 4 wavelet packet transform with:

```
> transfer.plot(wpt(1:512, n.level=4))
```

This is shown in figure 13.3. The wavelet packet bandpass filters have significant sidelobes. This causes the "shadows" which intersect the

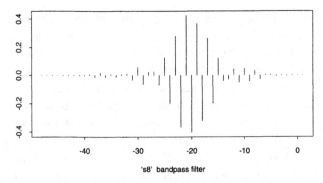

's8' bandpass filter

FIGURE 13.1. The wavelet packet filter corresponding to the filter sequence LLH.

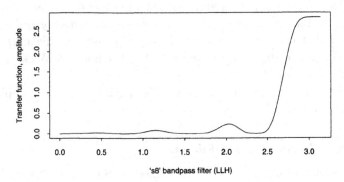

's8' bandpass filter (LLH)

FIGURE 13.2. The amplitude of the transfer function for the wavelet packet filter corresponding to the filter sequence LLH.

main line in the time-frequency plane plot for a linear chirp; see figure 7.10.

Note: The "data" 1:512 was used for convenience to create a wavelet packet transform for the transfer function plot. The transfer function depends only on the wavelet filters, and not on the data.

Figure 13.4 compares the transfer functions for the level 3 wavelet packet filters to those for the level 3 DWT filters. This plot is created as follows:

```
> par(mfrow=c(1,2))
> transfer.plot(wpt(1:512,n.level=3))
> transfer.plot(dwt(1:512,n.level=7))
```

The frequency bandwidth of the WPT filters is constant across all bands while the frequency bandwidth of the DWT filters gets smaller for coarser scales. Unlike the DWT, this type of WPT has the same frequency localization at all scales.

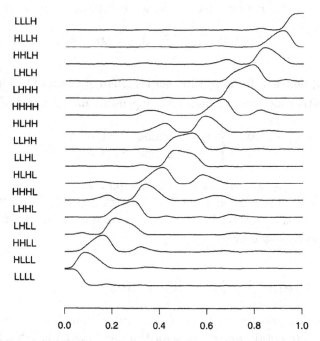

FIGURE 13.3. Transfer functions for a level 4 wavelet packet transform.

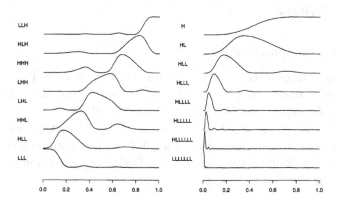

FIGURE 13.4. Transfer functions. Left: for a level 3 wavelet packet transform. Right: for the DWT.

13.1.3 Evaluating a Wavelet Packet Function

The dilation equation (section 12.5) lies at the heart of wavelet packet analysis as well as wavelet analysis. The dilation equation leads to a recursive formula for the wavelet packet functions:

$$W_b(x) = \sqrt{2} \sum_k f_k^{(b)} W_{\lfloor b/2 \rfloor}(2x - k) \, . \tag{13.1}$$

Here $\lfloor x \rfloor$ denotes the largest integer less than or equal to x. The filter $f_k^{(b)}$ is either low-pass or high-pass depending on the value of b:

$$f_k^{(b)} = \begin{cases} \ell_k & \text{if } b \bmod 4 = 0 \text{ or } 3 \\ h_k & \text{if } b \bmod 4 = 1 \text{ or } 2 \, . \end{cases}$$

The coefficients ℓ_k and h_k are the low-pass and high-pass wavelet filter coefficients defined in (12.1) and (12.2) respectively. Wavelet packet functions in S+WAVELETS are computed by the function `plot.wavelet.packet`.

13.2   Cosine Packet Algorithms

This section gives a brief overview of algorithms for cosine packet analysis. For details about the algorithms and methods behind cosine packet analysis (local cosine analysis), refer to the book by Wickerhauser [Wic94a].

13.2.1   Computing the DCT

The algorithms used to compute the DCT-II and DCT-IV transforms are summarized below. For details concerning algorithms for the DCT, refer to the book by Rao and Yip [RY90].

For a discrete signal $\mathbf{f} = (f_1, f_2, \ldots, f_n)$, the DCTs and their inverses can be computed from the FFT of the extended signal obtained by padding p zeros at the end. Define the extension operator $E_p(\mathbf{x})$ for the vector $\mathbf{x} = (x_1, x_2, \ldots, x_n)$ by

$$E_p(\mathbf{x}) = \begin{cases} x_t & t = 1, 2, \ldots, n \\ 0 & t = n+1, n+2, \ldots, n+p. \end{cases}$$

Define the FFT $G(\mathbf{y})$ of a vector $\mathbf{y} = (y_1, y_2, \ldots, y_m)$ by

$$G(\mathbf{y}) = \sum_{t=1}^{m} g_t \exp\left(-i2\pi k(t-1)/m\right) \qquad k = 0, 1, \ldots, m-1. \tag{13.2}$$

DCT-II

For the DCT-II, pad $p = n$ zeros, compute the FFT of the signal, and take the first n frequencies $\hat{f} = G(E_n(\mathbf{f}))$. The DCT-II coefficients $\mathbf{c} = (c_0, c_2, \ldots, c_{n-1})$ are given by

$$c_k = \sqrt{\frac{2}{n}} \left(\cos\left(\frac{k\pi}{2n}\right) \operatorname{Re}(\hat{f}_k) + \sin\left(\frac{k\pi}{2n}\right) \operatorname{Im}(\hat{f}_k) \right) . \qquad (13.3)$$

The inverse DCT-II is obtained by padding $p = 3n$ zeros to the DCT-II coefficients $\mathbf{c}$, dividing the first DCT coefficient by 2, and taking the FFT. From the FFT $\hat{c}_k$, the original signal is reconstructed as follows:

$$f_k = \sqrt{\frac{2}{n}} \operatorname{Re}(\hat{c}_{2k}) . \qquad (13.4)$$

For signals of length 2^J, the DCT-II is obtained from the discrete Hartley transform (DHT). This is a fast algorithm which avoids the need to double the signal length (as is the case when the DCT-II is computed from the FFT) [Mal86]. The discrete Hartley transform [SJBH85] is defined by

$$H_k = \sum_{t=1}^{n} \left(\cos\left(\frac{2k(t-1)\pi}{n}\right) + \sin\left(\frac{2k(t-1)\pi}{n}\right) \right) f_t \qquad (13.5)$$

for $k = 0, 1, \ldots, n-1$.

DCT-IV

For the DCT-IV, pad $p = 3n$ zeros and compute the FFT of the signal $\hat{f} = G(E_{3n}(\mathbf{f}))$. The DCT-IV coefficients are given by

$$c_k = \sqrt{\frac{2}{n}} \left(\cos\left(\frac{(k+.5)\pi}{2n}\right) \operatorname{Re}(\hat{f}_{2k+1}) + \sin\left(\frac{(k+.5)\pi}{2n}\right) \operatorname{Im}(\hat{f}_{2k+1}) \right)$$
$$(13.6)$$

The inverse DCT-IV is also given by (13.6), but with the coefficients and signal values reversed.

13.2.2 Computing the Cosine Packet Transform

Suppose you have partitioned a signal $\mathbf{f} = (f_1, f_2, \ldots, f_n)$ into P contiguous blocks $\mathbf{f}_1, \mathbf{f}_2, \ldots, \mathbf{f}_P$. The bth block has n_b coefficients and is given by

$$\mathbf{f}_b = (f_{i_b+1}, \ldots, f_{i_b+n_b}) = (f_{b,1}, \ldots, f_{b,n_b})$$

where $i_1 = 0$, $i_{b+1} = i_b + n_b$, $b = 1, \ldots, P$. Define a taper function B which applies a taper to each interval, as described in section 8.2.1. The same taper function is used for all intervals. Let $2m$ be the length of the taper; m corresponds to the optional argument n.taper. To apply the tapering formula (8.9), m must be less than or equal to half of the length of the *shortest* block.

The cosine packet transform is computed by applying five basic operations to each block $\mathbf{f}_b$:

1. Use the boundary rule to create boundary blocks $\mathbf{f}_0$ and $\mathbf{f}_{P+1}$ (see section 13.2.4 below).

2. Extend block $\mathbf{f}_b$ on the left and right using the neighboring blocks $\mathbf{f}_{b-1}$ and $\mathbf{f}_{b+1}$ to obtain the extended signal $\tilde{\mathbf{f}} = (\tilde{f}_{-m+1}, \tilde{f}_{-m+2}, \ldots, \tilde{f}_{n_b+m})$:

$$\tilde{f}_i = \begin{cases} f_{b-1,n_{b-1}+i} & \text{for } -m+1 \leq i \leq 0 \\ f_{b,i} & \text{for } 1 \leq i \leq n_b \\ f_{b+1,i} & \text{for } n_b+1 \leq i \leq n_b+m \, . \end{cases} \tag{13.7}$$

For a block of length n_b, the signal $\tilde{\mathbf{f}}$ has length $2m + n_b$.

3. Apply a taper $B(t)$ to the block $\tilde{\mathbf{f}}$ to obtain the tapered block $\mathbf{g}$ as follows:

$$g_i = \begin{cases} \sqrt{1 - B((i+.5)/m)^2}\,\tilde{f}_i & \text{for } -m+1 \leq i \leq 0 \\ B((i-.5)/m)\tilde{f}_i & \text{for } 1 \leq i \leq m \\ \tilde{f}_i & \text{for } m+1 \leq i \leq n_b-m \\ B((n_b-i+.5)/m)\tilde{f}_i & \text{for } n_b-m+1 \leq i \leq n_b \\ \sqrt{1 - B((i-n_b-.5)/m)^2}\,\tilde{f}_i & \text{for } n_b+1 \leq i \leq n_b+m \, . \end{cases} \tag{13.8}$$

See section 13.2.3 below for the tapering functions $B(t)$.

4. Fold m values on each end as follows:

$$\tilde{g}_i = \begin{cases} g_i + g_{-i+1} & \text{for } 1 \leq i \leq m \\ g_i & \text{for } m+1 \leq i \leq n_b-m \\ g_i - g_{n_b+(n_b-i)+1} & \text{for } n_b-m+1 \leq i \leq n_b \, . \end{cases} \tag{13.9}$$

This yields a "folded" signal $\tilde{\mathbf{g}} = (\tilde{g}_1, \tilde{g}_2, \ldots, \tilde{g}_{n_b})$ of length n_b. This folding is said to have $(+, -)$ polarity, since the reflected signal is added on the left boundary and subtracted on the right boundary.

5. Apply the DCT to the tapered and folded signal $\tilde{\mathbf{g}}$ to obtain the CPT coefficients $\mathbf{c}_b = (c_{b,1}, c_{b,2}, \ldots, c_{b,n})$.

The inverse cosine packet transform is obtained by reversing the above steps. Initialize the output signal $\mathbf{f}$ to a vector of n zeros $(0, 0, \ldots, 0)$. For each block of CPT coefficients $\mathbf{c}_b$, do the following six steps:

1. Apply the inverse DCT to the coefficients $\mathbf{c}_b$ to obtain the tapered and folded signal $\tilde{\mathbf{g}}$.

2. Unfold the signal using the $(+, -)$ polarity to obtain the unfolded signal $\mathbf{g}$ of length $n + 2m$:

$$
g_i = \begin{cases}
\tilde{g}_{-i+1} & \text{for } -m+1 \leq i \leq 0 \\
\tilde{g}_i & \text{for } 1 \leq i \leq n_b \\
-\tilde{g}_{n_b+(n_b-i+1)} & \text{for } n_b + 1 \leq i \leq n_b + m .
\end{cases}
\tag{13.10}
$$

3. Untaper $\mathbf{g}$ to obtain the extended signal $\tilde{\mathbf{f}}$:

$$
\tilde{f}_i = \begin{cases}
g_i / \sqrt{1 - B((i + .5)/m)^2} & \text{for } -m+1 \leq i \leq 0 \\
g_i / B((i - .5)/m) & \text{for } 1 \leq i \leq m \\
g_i & \text{for } m+1 \leq i \leq n_b - m \\
g_i / B((n_b - i + .5)/m) & \text{for } n_b - m + 1 \leq i \leq n_b \\
g_i / \sqrt{1 - B((n_b - i - .5)/m)^2} & \text{for } n_b + 1 \leq i \leq n_b + m.
\end{cases}
\tag{13.11}
$$

4. Create 3 vectors with lengths n_{b-1}, n_b, and n_{b+1}:

$$
\begin{aligned}
\tilde{f}_b^{(\ell)} &= (0, \ldots, 0, \tilde{f}_{-m+1}, \ldots, \tilde{f}_0) \\
\tilde{f}_b^{(c)} &= (\tilde{f}_1, \ldots, \tilde{f}_{n_b}) \\
\tilde{f}_b^{(r)} &= (\tilde{f}_{n_b+1}, \ldots, \tilde{f}_{n_b+m}, 0, \ldots, 0).
\end{aligned}
$$

5. The bth block is reconstructed by $f_b = \tilde{f}_{b-1}^{(r)} + \tilde{f}_b^{(c)} + \tilde{f}_{b+1}^{(\ell)}$.

6. Add the extended signal to the output signal $\mathbf{f}$:

$$
\begin{aligned}
f_{b-1, n_{b-1}+i} &= \tilde{f}_i & \text{for } -m+1 \leq i \leq 0 \\
f_{b,i} &= \tilde{f}_i & \text{for } 1 \leq i \leq n_b \\
f_{b+1, i-n_b} &= \tilde{f}_i & \text{for } n_b + 1 \leq i \leq n_b + m .
\end{aligned}
$$

Special treatment is needed for the blocks at the boundaries c_1 and c_P. See [AWW92, Wic94a] for more details about computing the CPT and the inverse CPT.

Note: The above algorithm applies to any set of blocks, not just dyadic ones. The dyadic blocking used in cosine packet analysis is convenient in constructing cosine packet tables.

13.2.3 Tapers for Orthogonal Cosine Packets

You can use any of 7 different tapers to do an orthogonal cosine packet analysis; see section 8.2.1. The specific form of these tapers is as follows:

boxcar   The boxcar tapering function is given by

$$B(t) = \begin{cases} 0 & \text{if } t \leq 0 \\ 1 & \text{if } 0 < t < 1 \\ 0 & \text{if } t \geq 1 \,. \end{cases}$$

poly1 , ..., poly5   A polynomial tapering function B of order $p = 1, \dots, 5$ is given by

$$B_p(t) = \sqrt{t^p \sum_{k=1}^{p} b_k t^{k-1}}, \qquad 0 \leq t \leq 1$$

where $(b_1, \dots, b_p)' = A_p^{-1}(1, 0, \dots, 0)'$ with

$$A_p = (a_{ij})_{1 \leq i,j \leq p}, \qquad a_{ij} = \begin{pmatrix} p+j-1 \\ i-1 \end{pmatrix}.$$

The polynomial tapers are available for $p = 1, 2, \dots, 5$, with higher order tapers providing a greater degree of smoothness.

trig   The tapering function for the trigonometric window is given by

$$B(t) = \sin\left(\frac{\pi}{4}\left(1 - \cos(\pi t)\right)\right) \qquad 0 \leq t \leq 1.$$

The trigonometric taper is the smoothest taper available in S+WAVELETS.

13.2.4 Boundary Extension Rules

Since the tapers extend beyond the ends of the analysis intervals, it is necessary to extend the data at boundaries in some manner. This is similar to the boundary correction algorithms required in wavelet analysis (see chapter 14). You can select a boundary extension in the functions `block.cpt`, `cp.table`, `cpt`, `cpt.2d`, and `cp.costs.2d` using the optional argument `boundary`.

For a taper of length $2m \leq n$, it is necessary to extend a signal $(f_1, f_2, \ldots, f_n)$ of length n by m values on each side to a obtain a signal of length $n + 2m$ $(\tilde{f}_{-m+1}, \tilde{f}_{-m+2}, \ldots, \tilde{f}_{n+m})$. There are three boundary extension rules available in S+WAVELETS:

`cp.reflect:` The signal is reflected at the ends using the same $(+, -)$ polarity as the folding operator. The extended signal is

$$\tilde{f}_i = \begin{cases} f_{-i+1} & \text{for } -m+1 \leq i \leq 0 \\ f_i & \text{for } 1 \leq i \leq n \\ -f_{2n-i+1} & \text{for } n+1 \leq i \leq n+m \ . \end{cases}$$

`zero:` The signal is zero padded, so the extended signal is

$$\tilde{f}_i = \begin{cases} 0 & \text{for } -m+1 \leq i \leq 0 \\ f_i & \text{for } 1 \leq i \leq n \\ 0 & \text{for } n+1 \leq i \leq n+m \ . \end{cases}$$

`periodic:` The signal is assumed to be periodic and wrapped around at the ends, so the extended signal is

$$\tilde{f}_i = \begin{cases} f_{n-i} & \text{for } -m+1 \leq i \leq 0 \\ f_i & \text{for } 1 \leq i \leq n \\ f_{i-n} & \text{for } n+1 \leq i \leq n+m \ . \end{cases}$$

This is the default boundary extension rule in S+WAVELETS for cosine packet analysis.

Only the periodic boundary extension preserves orthogonality for the boundary blocks. The boundary blocks for the `cp.reflect` and `zero` extensions are only nearly orthogonal. As an example, we compute the correlation between the signal components for the "level 3" smooth CPT of a linear chirp signal:

```
> lc <- make.signal("linchirp", n=1024)
> lc.cpt3 <- block.cpt(lc, n.level=3, boundary="zero")
> summary(decompose(lc.cpt3, order="time"))
```

Correlation Matrix:

	Data	C3.0	C3.1	C3.2	C3.3	C3.4	C3.5	C3.6
	1.000	0.367	0.359	0.359	0.359	0.358	0.358	0.357
C3.0	0.367	1.000	0.002	0.000	0.000	0.000	0.000	0.000
C3.1	0.359	0.002	1.000	0.000	0.000	0.000	0.000	0.000
C3.2	0.359	0.000	0.000	1.000	0.000	0.000	0.000	0.000
C3.3	0.359	0.000	0.000	0.000	1.000	0.000	0.000	0.000
C3.4	0.358	0.000	0.000	0.000	0.000	1.000	0.000	0.000
C3.5	0.358	0.000	0.000	0.000	0.000	0.000	1.000	0.000
C3.6	0.357	0.000	0.000	0.000	0.000	0.000	0.000	1.000
C3.7	0.309	0.001	0.000	0.000	0.000	0.000	0.000	0.000

	C3.7
	0.309
C3.0	0.001
C3.1	0.000
C3.2	0.000
C3.3	0.000
C3.4	0.000
C3.5	0.000
C3.6	0.000
C3.7	1.000

Variances:

Data	C3.0	C3.1	C3.2	C3.3	C3.4	C3.5	C3.6	C3.7
0.489	0.065	0.063	0.063	0.063	0.063	0.063	0.062	0.046

Statistics for Components:

	Min	1Q	Median	3Q	Max	Mean	SD	MAD	Energy.%
C3.0	-1.150	0	0	0	0.999	0.036	0.256	0	0.136
C3.1	-1.081	0	0	0	1.181	-0.004	0.251	0	0.128
C3.2	-1.184	0	0	0	1.138	-0.001	0.251	0	0.129
C3.3	-1.158	0	0	0	1.183	0.000	0.251	0	0.128
C3.4	-1.181	0	0	0	1.181	0.000	0.250	0	0.128
C3.5	-1.174	0	0	0	1.200	0.000	0.250	0	0.128
C3.6	-1.190	0	0	0	1.196	0.000	0.250	0	0.128
C3.7	-1.000	0	0	0	1.000	-0.001	0.216	0	0.095

The correlation between the left and right boundary blocks $C_{3,0}$ and $C_{3,7}$ with other blocks is at most 0.002. The interior blocks are orthogonal for all boundary extension rules.

Because the transform is not exactly orthogonal, the total energy of the transform may be different from the total energy of the signal for the `cp.reflect` and `zero` extensions.

13.3   Algorithms for 2-D Packets

The algorithms for 2-D wavelet packets and cosine packets are quite straightforward adaptations of their 1-D counterparts. For details about algorithms for 2-D wavelet packet analysis and cosine packet analysis (local cosine analysis), refer Wickerhauser [Wic94a].

13.3.1   The 2-D WPT Algorithm

The algorithm for computing a 2-D wavelet packet transform (WPT) is a straightforward generalization of the 2-D DWT pyramid algorithm, just as the 1-D WPT is a generalization of the 1-D DWT pyramid algorithm. See section 12.6 for details concerning the 2-D pyramid algorithm, and section 13.1 regarding the 1-D wavelet packet algorithm.

The 2-D WPT algorithm proceeds by recursively filtering coefficients blocks, starting with the data. Like the 2-D DWT pyramid algorithm, at each filtering stage a coefficient block is filtered and down-sampled to create four subblocks; see figure 12.11. In contrast to the 2-D DWT, filtering is applied to other blocks beside the smooth (i.e., oscillation 0) coefficient blocks.

The 2-D WPT algorithm is illustrated using a level 2 WPT. First, obtain the "internal" crystal names for a level 2 WPT using the function `icrystal.names`:

```
> wpt2d.nms <- icrystal.names("subband.2d", J=2)
> wpt2d.nms
 [1] "LL-LL" "HL-LL" "HH-LL" "LH-LL"
 [5] "LL-HL" "HL-HL" "HH-HL" "LH-HL"
 [9] "LL-HH" "HL-HH" "HH-HH" "LH-HH"
[13] "LL-LH" "HL-LH" "HH-LH" "LH-LH"
attr(, "is.on"):
 orthogonal complete
         T        T
```

The internal crystal names use the filtering name convention described in section 5.2. Internal names are used for all of the internal calculations in S+WAVELETS functions. The arguments `subband.2d` and `J=2` generate all wavelet packet crystal names at level 2.

Use the function **get.parents.quad** to find out the sequence of "parent" crystals which must be filtered to obtain a level 2 WPT:

```
> wpt2d.parents <- get.parents.quad(wpt2d.nms)
> wpt2d.parents
     levels row-blocks col-blocks
  -       0          0          0
L-L       1          0          0
L-H       1          0          1
H-L       1          1          0
H-H       1          1          1
```

The 2-D WPT proceeds by filtering the crystals in **wpt2d.parents** starting from the first row and continuing downwards.

A level J wavelet packet cost table is obtained by simply computing a level J wavelet packet transform. The costs of all crystals are saved at levels $j = 0, 1, \ldots, J$ in going from level 0 to level J.

13.3.2   The 2-D CPT Algorithm

The 2-D DCT and 2-D CPT algorithms are a simple generalizations of the 1-D cases.

The 2-D DCT is obtained by first computing the 1-D DCT across rows of an image $\mathbf{X}$ to obtain the row-transformed image $\mathbf{X}_r$. The 1-D DCT is then applied across columns to obtain the 2-D DCT transform image $\mathbf{X}_{\text{dct}}$. The row and column operations can be permuted. Turn to section 13.2.1 for a description of the 1-D DCT algorithms.

To compute the 2-D CPT, start by partitioning the image $\mathbf{X}$ into P contiguous and non-overlapping blocks $\mathbf{X}_1, \mathbf{X}_2, \ldots, \mathbf{X}_P$. The 2-D CPT is computed by first tapering and folding blocks $\mathbf{X}_1, \mathbf{X}_2, \ldots, \mathbf{X}_P$ as follows:

1. Extend the block $\mathbf{X}_i$ across rows by m values at each end to obtain the row extended block $\widetilde{\mathbf{X}}_i^r$.

2. Apply a taper $B(t)$ across rows of block $\widetilde{\mathbf{X}}_i^r$ to obtain the tapered block $\mathbf{Y}_i^r$.

3. Fold the tapered block $\mathbf{Y}_i^r$ across rows using $(+, -)$ polarity to obtain a folded tapered block $\widetilde{\mathbf{Y}}_i^r$. The folded block has the same dimensions as the original block $\mathbf{X}_i$.

4. Extend the block $\widetilde{\mathbf{Y}}_i^r$ across columns by m values at each end to obtain the column extended block $\widetilde{\mathbf{X}}_i$.

5. Apply a taper $B(t)$ across columns of block $\widetilde{\mathbf{X}}_i$ to obtain the tapered block $\mathbf{Y}_i$.

6. Fold the tapered block $\mathbf{Y}_i$ across columns using $(+, -)$ polarity to obtain a folded tapered block $\widetilde{\mathbf{Y}}_i$. Replace the original block $\mathbf{X}_i$ by the folded tapered block $\widetilde{\mathbf{Y}}_i$.

The 2-D CPT is computed by applying the 2-D DCT to each of the folded tapered blocks $\widetilde{\mathbf{Y}}_i$ for $i = 1, 2, \ldots, P$.

The 2-D extension, tapering, and folding operations are identical to their 1-D counterparts. Turn to section 13.2 for details about these operations.

14
Boundary Conditions for Wavelet Analysis

The "theoretical" wavelet transform operator $\mathbf{W}^*$ maps an infinite signal $f(t)$ to an infinite set of wavelet coefficients. To apply $\mathbf{W}^*$ to a finite signal, we need to truncate $\mathbf{W}^*$ to a finite invertible operator $\mathbf{W}$ or extend the finite signal to an infinite one. There are many ways to do the truncation or extension. The resulting transform is the same for the "interior" coefficients; i.e., coefficients far enough away from the boundaries. What differs are the coefficients produced at the boundaries.

In some cases, changing the default boundary treatment may be appropriate. In this chapter, you will learn about the following topics:

- Which boundary treatment rules are available (section 14.1).

- Considerations in the selection of a boundary treatment rule for wavelet analysis (section 14.2).

- Comparison of boundary treatment rules for wavelet analysis (section 14.3).

- Algorithms for adapting the wavelet filters at the boundaries (section 14.4).

14.1   Types of Boundary Treatment Rules

The wavelet analysis functions dwt, mrd, mra, wp.table, dwt.2d, wpt.2d, and wp.costs.2d all take a boundary argument. There are seven types of boundary treatment options for wavelet analysis supported in S+WAVELETS:

periodic

The original series $f_1, f_2, \ldots, f_n$ is assumed to be n periodic: $f_{i+nk} = f_i$ for $i = 1, 2, \ldots, n$ and $k = \ldots, -1, 0, 1, \ldots$. Equivalently, the wavelets are assumed to be periodic on the interval $[0, n]$.

reflection

The original series $f_1, f_2, \ldots, f_n$ is reflected at the boundaries and then periodically extended using the algorithm given by [Bri92]. To ensure perfect reconstruction, the reflection boundary correction is only available for symmetric wavelets; i.e., the biorthogonal wavelets.

zero

At each filtering step, the coefficients are padded at the beginning and the end of the signal with zeros.

poly0

At each filtering step, the coefficients are padded at the beginning and the end of the signal by repeating the first and last value respectively.

poly1

At each filtering step, the coefficients are padded at the beginning and the end of the signal using a polynomial extension of degree one, fit to the first two and last two values respectively.

poly2

At each filtering step, the coefficients are padded at the beginning and the end of the signal using a polynomial extension of degree two, fit to the first three and last three values respectively.

interval

This rule corresponds to the special wavelet functions at the boundaries defined by Cohen, Daubechies and Vial [CDV93]. The boundary wavelets are zero outside of the range of the data. The transform retains the orthogonality properties of the "classical" wavelet transform and is numerically stable.

For the interval wavelet, one can specify an additional argument called **precondition**. If **precondition=T**, then a preconditioning transform is applied to the signal before applying the interval wavelet transform. The preconditioning transform preserves the "vanishing moment" property of wavelets (see section 12.4) at the expense of introducing an additional non-orthogonal transform.

infinite

The original series $f_1, f_2, \ldots, f_n$ is extended once at the beginning of the filtering procedure using a zero or polynomial extension. This is similar to **zero** and **poly0-poly2** options, for which the coefficients are extended at each filter step. The polynomials are fit using a fraction **pfrac** of the data ($0 \leq$ **pfrac** < 1). The argument **pdeg** specifies the degree of the polynomial extension (**pdeg** ≥ 0) or a zero extension (**pdeg** $= -1$).

The **infinite** boundary rule produces an infinite set of wavelet coefficients. However, only a finite number of coefficients is stored. Since a polynomial extension rule is used, the stored coefficients can be used to compute the remaining coefficients.

Unlike the other options, the **infinite** boundary treatment is an "expansionist" transform; it generates more coefficients than original sample values. For a series of length n, you obtain $n + p$ wavelet coefficients where $0 < p \ll n$, independent of n. As a result,

Boundary Condition	Wavelet Type	Transform Type	Sample Size
periodic	any	any	2^J
poly0	any	any	any
poly1	any[1]	any	any
poly2	any[2]	any	any
zero	any	any	any
reflection	any[3]	any	any
interval	s4-s16	dwt, dwt.2d	2^J
infinite	orthogonal	any 1-D	any

[1] The poly1 boundary rule is not available with wavelets bs2.2, bs2.4, bs2.6, and bs2.8 using the dual=T option.

[2] The poly2 boundary rule is not available with wavelets bs3.1, bs3.3, bs3.5, bs3.7, and bs3.9 using the dual=T option.

[3] For the reflection boundary rule, you get perfect reconstuction only with the biorthogonal symmetric/anti-symmetric wavelets.

TABLE 14.1. Restrictions on sample sizes, wavelet types, and transforms for various boundary conditions. "2^J" means the length of the series must be divisible 2^J where J is the maximum number of levels (n.level). "Any 1-D" means that any 1-D transform can be applied.

a DWT object computed using the infinite boundary condition is a "crystal list" object rather than a crystal vector object (see appendix C).

The boundary treatments are not uniformly available for all types of wavelets, transforms, and sample sizes. Table 14.1 summarizes which boundary options can be used for a given sample size and wavelet filter.

The default boundary treatment rules are:

- periodic for sample sizes divisible by 2^J where J is the maximum number of levels.

- zero for sample sizes not divisible by 2^J when using orthogonal wavelets.

- reflection for sample sizes not divisible by 2^J when using biorthogonal wavelets.

The periodic boundary condition is available only for sample sizes divisible by 2^J. For sample sizes not divisible by 2^J, you must use a different boundary condition.

The **periodic** and **reflection** boundary conditions are both the fastest and the numerically most stable transforms and inverse transforms.

 Warning: The **periodic** boundary condition can lead to boundary artifacts in certain signals and images since it assumes the signal or image is periodic. In this case, you may want to use a different boundary rule.

 Warning: The **reflection** boundary condition gives perfect reconstruction only for biorthogonal symmetric/anti-symmetric filters.

 Warning: The **poly0-poly2** methods are numerically less accurate for very long wavelet filters (**d18**, **d20**, **s18**, **s20**, and **c30**). For these wavelets, you should use either the **periodic** or **zero** boundary methods to ensure accurate inverse transform methods.

14.2 Considerations in Selecting a Boundary Method

In choosing a boundary treatment, you must balance a number of considerations. A good choice will do all of the following:

1. *Ensure perfect reconstruction of the original signal.* You may not get perfect reconstruction of the original signal or image with the **reflection** treatment rule when using orthogonal wavelets. With biorthogonal wavelets, the **reflection** treatment rule does lead to perfect reconstruction.

2. *Match properties of the signal.* Based on known properties of a signal, you may have reason to prefer a given boundary treatment method. If your signal is periodic outside the sampling interval, then you should use the **periodic** boundary method. Otherwise, zero padding may be appropriate and you should use the **zero** boundary method.

3. *Be fast computationally.* The **periodic** and **reflection** treatment rules are the fastest (**periodic** is available only for sample sizes divisible by 2^J). The **infinite** and **interval** boundary conditions are not coded in C. For the functions **dwt**, **mrd**, and **mra**, this makes only a moderate difference. However, the func-

tions wp.table (for infinite) and dwt.2d (for interval) are
very slow using these boundary conditions.

4. *Minimize artifacts.* An important criterion is to avoid creating large "artificial" coefficients at the boundaries. For most
applications, the best boundary conditions in this regard are
reflection and interval. The boundary artifacts are especially prominent for wavelet packet analysis.

5. *Apply to all types of wavelets and signal lengths.* You want a
boundary treatment operator which can work for a variety of
wavelets and sample sizes. The drawback of interval wavelets
is the restriction on their use to symmlets, samples of size 2^J,
and only for the DWT.

6. *Lead to a non-expansionist transform.* A non-expansionist transform means that if you start with n sampled values, you have n
wavelet coefficients. Expansionist transforms lead to more than
n wavelet coefficients. In S+WAVELETS the only expansionist
transform is infinite.

7. *Preserve vanishing moments property.* A mother wavelet ψ on
the real line with M vanishing moments satisfies

$$\int x^m \psi(x) dx = 0 \qquad m = 0, 1, \ldots, M - 1.$$

If the level 0 wavelet coefficients $s_{0,k}$ are polynomial of degree
$M - 1$, then

- the coefficients $s_{1,k}$ derived from the low-pass filter are also
  a polynomial of degree $M - 1$,
- the level 1 coefficients $d_{1,k}$ derived from the high-pass filter
  are identically zero.

The poly0-poly2 boundary treatment will preserve polynomials up to degree 0-2 respectively. If precondition=T, the
interval treatment will preserve polynomials up to degree of
the wavelet.

8. *Preserve orthogonality.* Ideally, the discrete wavelet transform
operator should be orthogonal. The interval and periodic
boundary treatments preserve orthogonality. The reflection
boundary treatment preserves biorthogonality.

9. *Have good numerical accuracy.* The inverse transform applied to transformed data should return the original data. The inverted transform will differ from the original data by a small number due to round-off error, numerical accuracy of the filters, and the stability of the boundary treatment method. The `periodic` and `reflection` are the most stable boundary methods. The least stable methods are `poly0-poly2` and `interval`. The `poly0-poly2` methods are not recommended for very long wavelet filters such as d18, d20, s18, s20, and c30.

For sample sizes divisible by 2^J where J is the maximum number of levels in the transform, the `periodic` boundary condition is fast and works for both orthogonal and biorthogonal wavelets. This is why, for these sample sizes, the default boundary condition is `periodic`.

For biorthogonal wavelets, a good alternative to the `periodic` boundary rule is the `reflection` boundary rule, which may reduce artifacts at the boundaries. Another good choice is the `interval` wavelet, although this is available only for symmlets, and for series of length divisible by 2^J where J is the maximum number of levels in the transform. In addition, the `interval` wavelet is restricted to the discrete wavelet transform and is computationally slower.

The `infinite` boundary option is not recommended for general use, but is available for researchers and special applications. This method is computationally slow and leads to an expansionist transform.

14.3 Comparing Boundary Treatments

To get an understanding of the effect of the boundary treatments on the coefficients, look at the DWT of sampled values of simple functions. This S-PLUS code compares the DWT for several different boundary options, for a linear function:

```
> m1 <- as.matrix(dwt(1:10, wavelet="s4",
+     boundary="poly0", n.level=1))
> m2 <- as.matrix(dwt(1:10, wavelet="s4",
+     boundary="poly1", n.level=1))
> m3 <- as.matrix(dwt(1:10, wavelet="s4",
+     boundary="zero", n.level=1))
> m4 <- as.matrix(dwt(1:10, wavelet="s4",
+     boundary="interval", n.level=1))
```

```
> m5 <- as.matrix(dwt(1:10, wavelet="s4",
+     boundary="interval", n.level=1, precondition=T))
> boundaries <- c("Poly 0", "Poly 1", "Zero", "Interval",
+   "Precond")
> mat <- cbind(m1, m2, m3, m4, m5)
> dimnames(mat) <- list(dimnames(mat)[[1]], boundaries)
> round(mat,2)
      Poly 0 Poly 1   Zero Interval Precond
s1(1)   3.22   3.35   3.35     2.43    2.07
s1(2)   6.17   6.17   6.17     6.60    6.73
s1(3)   9.00   9.00   9.00     9.00    9.00
s1(4)  11.83  11.83  11.83    12.08   11.85
s1(5)  14.18  14.66   9.35     9.06    5.20
d1(1)  -0.48   0.00   0.00    -0.27    0.00
d1(2)   0.00   0.00   0.00     0.25    0.00
d1(3)   0.00   0.00   0.00     0.00    0.00
d1(4)   0.00   0.00   0.00     0.06    0.00
d1(5)   0.13   0.00   1.42    -5.12    0.00
```

The "best" boundary corrections for a linear function are the polynomial extension of degree 1 and the preconditioned interval wavelet transform. For both of these boundary corrections, the detail wavelet coefficients are zero and a linear function is entirely represented by the smooth wavelet coefficients. The detail coefficients are relatively small for the polynomial extension of degree 0 and for the interval wavelet transform without preconditioning. However, the zero extension exhibits serious artifacts; this transform has large detail coefficients which are entirely due to the boundary correction method.

Use one of the biorthogonal symmetric filters in order to look at the reflection wavelet operator. As an example, compare the effects of reflection and periodic boundary treatments for the bs2.2 filter with a linear function as follows:

```
> m1 <- dwt(1:8, wavelet="bs2.2", boundary="periodic",
+     n.level=1)
> m2 <- dwt(1:8, wavelet="bs2.2", boundary="reflection",
+     n.level=1)
> mat <- cbind(as.matrix(m1), as.matrix(m2))
> dimnames(mat) <- list(dimnames(as.matrix(m1))[[1]],
+       c("Periodic", "Reflection"))
> round(mat, 2)
      Periodic Reflection
s1(1)     4.24       2.12
```

s1(2)	4.24	4.24
s1(3)	7.07	7.07
s1(4)	9.90	9.90
d1(1)	−1.41	−0.35
d1(2)	0.00	0.00
d1(3)	0.00	0.00
d1(4)	4.24	1.41

The detail coefficients for the reflection boundary treatment are small but non-zero, due to the change in slope at the boundaries of the reflected series. By contrast, the detail coefficients for the periodic boundary treatment are large since a discontinuity is created in the extended series.

14.4   Boundary Correction Algorithms

To implement the discrete wavelet transform using the pyramid algorithm, it is necessary to adapt the wavelet filters at the ends of the series. The way the filters are adapted depends on the boundary correction method selected; see section 14.1. In general, the algorithms for implementing the different boundary treatments are quite involved.

In S+WAVELETS, you can obtain a precise algorithmic description for each of the boundary correction methods by examining the appropriate convdown and convup functions. For example, the periodic boundary correction is implemented by the convdown.periodic function:

```
> convdown.periodic
function(x, f)
{
x <- unclass(x)
n <- length(x)
p <- length(f)/2 - 1
if(p > 0)
x1 <- c(x[(n - p + 1):n], x, x[1:p])
else x1 <- x
convdown.general(x1, f)
}
```

The input vector x is extended by p = L/2 − 1 values on each side where L is the length of the filter f. The extension is done by wrapping the vector around. The function convdown.general performs

Functions	Boundary Correction Method
convdown.polynomial convup.polynomial	poly0, poly1, poly2, zero
convdown.reflection convup.reflection	reflection
convdown.interval convup.interval	interval
convdown.periodic convup.periodic	periodic
convdown.infinite convup.infinite	infinite

TABLE 14.2. Table of functions implementing the convup and convdown operators for different boundary correction methods.

the filtering and down-sampling operation. It takes as input the extended data x1 of length n + 2p and outputs a vector of length n, where n is the length of the input vector x. See table 14.2 for a list of functions and corresponding boundary treatment methods.

Note: With the exception of the functions convdown.infinite and convdown.interval, the convdown and convup operators are not used to compute the transforms. For efficiency, most transforms are computed entirely in C.

14.4.1 Examining the Matrix Operators

Use the function dwt.matrix to obtain the wavelet operators $\mathbf{W}_1$ for different boundary treatments with n.level=1. For example, we obtain the periodic wavelet operator using the "periodic" boundary method with the symmlet "s4" as follows:

```
> round(dwt.matrix(wavelet="s4", n=8,
+    boundary="periodic"), 2)
        [,1]  [,2]  [,3]  [,4]  [,5]  [,6]  [,7]  [,8]
1(1)   0.22  0.84  0.48  0.00  0.00  0.00  0.00 -0.13
1(2)   0.00 -0.13  0.22  0.84  0.48  0.00  0.00  0.00
1(3)   0.00  0.00  0.00 -0.13  0.22  0.84  0.48  0.00
1(4)   0.48  0.00  0.00  0.00  0.00 -0.13  0.22  0.84
h(1)   0.84 -0.22 -0.13  0.00  0.00  0.00  0.00 -0.48
h(2)   0.00 -0.48  0.84 -0.22 -0.13  0.00  0.00  0.00
h(3)   0.00  0.00  0.00 -0.48  0.84 -0.22 -0.13  0.00
h(4)  -0.13  0.00  0.00  0.00  0.00 -0.48  0.84 -0.22
```

The first four rows produce coefficients s_1 and the last four rows produce coefficients d_1. See how the wavelet filters wrap around at the ends. Compare with the "zero" wavelet operator:

```
> round(dwt.matrix(wavelet="s4", n=8, boundary="zero"),2)
       [,1]   [,2]   [,3]   [,4]   [,5]   [,6]   [,7]   [,8]
1(1)  0.22   0.84   0.48   0.00   0.00   0.00   0.00   0.00
1(2)  0.00  -0.13   0.22   0.84   0.48   0.00   0.00   0.00
1(3)  0.00   0.00   0.00  -0.13   0.22   0.84   0.48   0.00
1(4)  0.00   0.00   0.00   0.00   0.00  -0.13   0.22   0.84
h(1)  0.84  -0.22  -0.13   0.00   0.00   0.00   0.00   0.00
h(2)  0.00  -0.48   0.84  -0.22  -0.13   0.00   0.00   0.00
h(3)  0.00   0.00   0.00  -0.48   0.84  -0.22  -0.13   0.00
h(4)  0.00   0.00   0.00   0.00   0.00  -0.48   0.84  -0.22
```

Appendix A
Orthogonal Wavelet Functions

This appendix is a catalog of the orthogonal wavelets available in S+WAVELETS. The *daublets* are plotted in figures A.1 and A.2, the *coiflets* are plotted in figure A.3, and the *symmlets* are plotted in figures A.4-A.5. A number of properties about wavelet functions (the support width, the number of vanishing moments, the number of derivatives, and the approximate Hölder exponent) are summarized in table A.1.

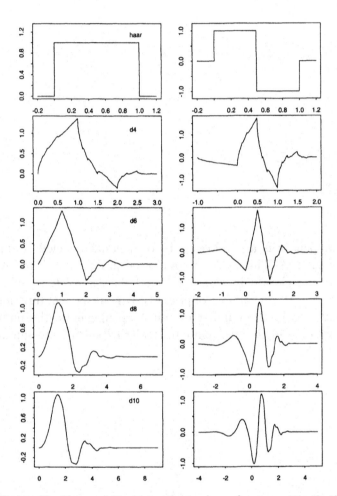

FIGURE A.1. The Haar and Daublet orthogonal wavelets **haar**, **d4**, **d6**, **d8**, and **d10**. Left: father wavelets. Right: mother wavelets.

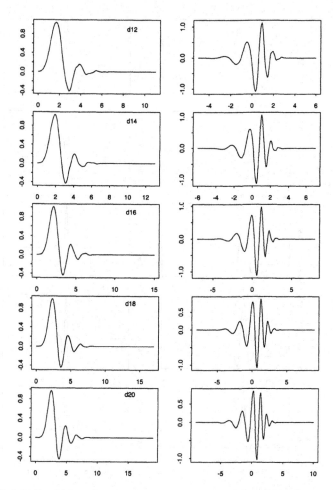

FIGURE A.2. Daublet orthogonal wavelets d12, d14, d16, d18, and d20. Left: father wavelets. Right: mother wavelets.

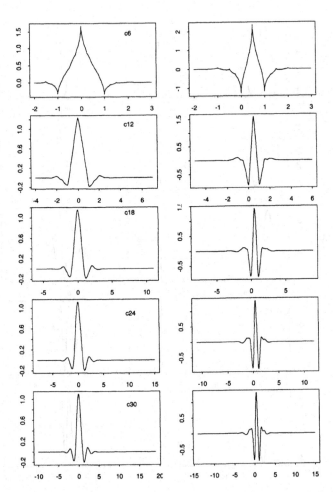

FIGURE A.3. Coiflet orthogonal wavelets c6, c12, c18, c24, and c30. Left: father wavelets. Right: mother wavelets.

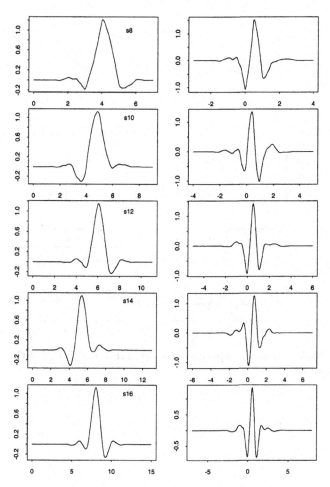

FIGURE A.4. Symmlet orthogonal wavelets s8, s10, s12, s14, and s16. Left: father wavelets. Right: mother wavelets.

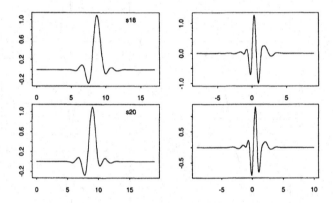

FIGURE A.5. Symmlet orthogonal wavelets s18 and s20. Left: father wavelets. Right: mother wavelets.

Wavelet	Support Length ψ, ϕ	Vanishing Moments ψ	Vanishing Moments ϕ	Number of Derivatives ψ, ϕ	Hölder Exponent daublets/symmlets ψ, ϕ
haar	1	0	0	0	0
d4	3	1	0	0	0.55
d6	5	2	0	1	1.09
d8, s8	7	3	0	1	1.62/1.40
d10, s10	9	4	0	1	1.97/1.78
d12, s12	11	5	0	2	2.19/2.12
d14, s14	13	6	0	2	2.46/2.47
d16, s16	15	7	0	2	2.76/2.75
d18, s18	17	8	0	3	3.07/3.04
d20, s20	19	9	0	3	3.38/3.31
c6	5	1	2	0	0.54
c12	11	3	4	1	1.45
c18	17	5	6	2	2.21
c24	23	7	8	2	2.87
c30	29	9	10	3	3.47

TABLE A.1. Table of orthogonal wavelets and their properties.

Table A.1 summarizes the support width, the number of vanishing moments, the number of derivatives, and the approximate Hölder exponent for the orthogonal wavelets.

Refer to (12.11) in section 12.4 for the definition of the number of vanishing moments. The Hölder exponent, which measures smoothness, is defined by $\alpha = m + \beta$ where α and β are the largest values such that

$$|f^{(m)}(x) - f^{(m)}(x + t)| \leq C|t|^{\beta} \qquad \text{for all } x, t. \qquad (A.1)$$

Here $f^{(m)}$ is the mth derivative of the wavelet f.

The *daublets* and *symmlets* have the same support width, number of vanishing moments, and number of derivatives, and nearly the same Hölder exponent. The main difference between *symmlets* and *daublets* is that *symmlets* are nearly symmetric while *daublets* are highly asymmetric.

Note: The Hölder exponents are derived using the techniques described in Chapter 7 of [Dau92].

Appendix B
Biorthogonal Wavelet Functions

This appendix is a catalog of the orthogonal wavelets included with S+WAVELETS. The *b-spline* wavelets are plotted in figures B.1-B.6 and the *v-spline* wavelets are plotted in figures B.7-B.8. A number of properties of biorthogonal wavelet functions are summarized in table B.1—the support width, the number of vanishing moments, the number of derivatives, and continuity.

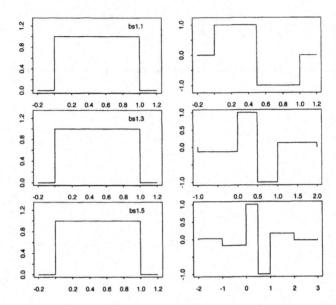

FIGURE B.1. Biorthogonal b-spline wavelets **bs1.1**, **bs1.3**, and **bs1.5**. Left: father wavelets. Right: mother wavelets. The corresponding dual wavelets are plotted in figure B.2.

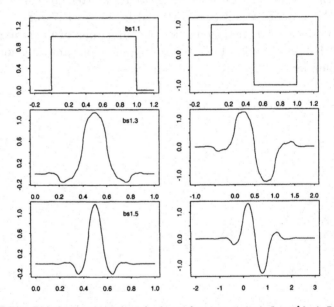

FIGURE B.2. Biorthogonal b-spline dual wavelets **bs1.1**, **bs1.3**, and **bs1.5**. Left: father wavelets. Right: mother wavelets. The corresponding wavelets are plotted in figure B.1.

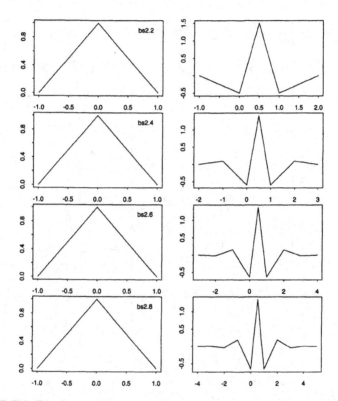

FIGURE B.3. Biorthogonal b-spline wavelets **bs2.2**, **bs2.4**, **bs2.6**, and **bs2.8**. Left: father wavelets. Right: mother wavelets. The corresponding dual wavelets are plotted in figure B.4.

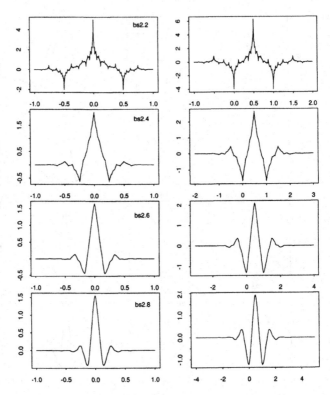

FIGURE B.4. Biorthogonal b-spline dual wavelets bs2.2, bs2.4, bs2.6, and bs2.8. Left: father wavelets. Right: mother wavelets. The corresponding wavelets are plotted in figure B.3.

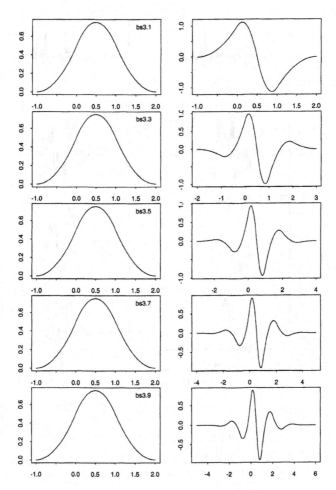

FIGURE B.5. Biorthogonal b-spline wavelets **bs3.1**, **bs3.3**, **bs3.5**, **bs3.7**, and **bs3.9**. Left: father wavelets. Right: mother wavelets. The corresponding dual wavelets are plotted in figure B.6.

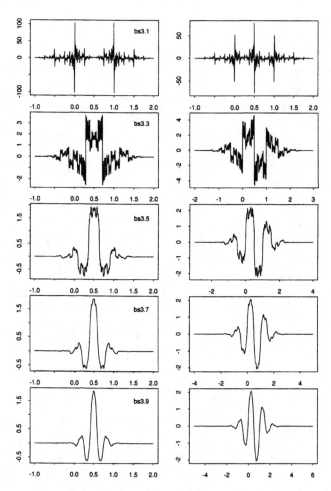

FIGURE B.6. Biorthogonal b-spline dual wavelets bs3.1, bs3.3, bs3.5, bs3.7, and bs3.9. Left: father wavelets. Right: mother wavelets. The corresponding wavelets are plotted in figure B.5.

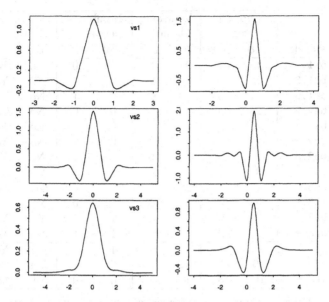

FIGURE B.7. Biorthogonal b-spline wavelets vs1, vs2, and vs3. Left: father wavelets. Right: mother wavelets. The corresponding dual wavelets are plotted in figure B.8.

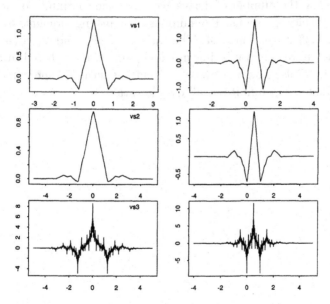

FIGURE B.8. Biorthogonal b-spline dual wavelets vs1, vs2, and vs3. Left: father wavelets. Right: mother wavelets. The corresponding wavelets are plotted in figure B.7.

Wavelet	Support Length $\phi, \tilde{\phi}$	Support Length $\psi, \tilde{\psi}$	Vanishing Moments ψ	Vanishing Moments $\tilde{\psi}$	Number of Derivatives ϕ, ψ	Number of Derivatives $\tilde{\phi}, \tilde{\psi}$	Continuous ϕ, ψ	Continuous $\tilde{\phi}, \tilde{\psi}$
bs1.1	1	1	0	0	0	0	No	No
bs1.3	1	5	2	4	0	1	No	Yes
bs1.5	1	9	4	8	0	1	No	Yes
bs2.2	2	4	2	4	0	0	No	No*
bs2.4	2	8	4	8	0	0	No	No
bs2.6	2	12	6	12	0	1	No	Yes
bs2.8	2	16	8	16	0	2	No	Yes
bs3.1	3	3	0	2	1	0	Yes	No*
bs3.3	3	7	2	6	1	0	Yes	No*
bs3.5	3	11	4	10	1	0	Yes	No
bs3.7	3	15	6	14	1	1	Yes	Yes
bs3.9	3	19	8	18	1	1	Yes	Yes
vs1	6	8	5	5	1	1	Yes	Yes
vs2	10	8	5	5	2	1	Yes	Yes
vs3	10	8	5	5	3	0	Yes	No*

TABLE B.1. Table of biorthogonal wavelets and their properties.

Table B.1 summarizes the support width, the number of vanishing moments, the number of derivatives, and the continuity of biorthogonal wavelet functions. The number of vanishing moments is defined by (12.11) in section 12.4. The father wavelet and the dual father wavelet have zero vanishing moments, and are not listed in column 2 of the table. The dual wavelet functions with "*" indicated under Continuous are infinite at all dyadic rationals.

Appendix C
Wavelet Classes and Objects

This appendix discusses all classes and objects that are defined in S+WAVELETS. The objects are divided into the following groups:

- 1-D transform, decompose and other analysis objects (see section C.1).

- Crystals, molecules, and atoms (see section C.2).

- 2-D transform and analysis objects (see section C.3).

- Wavelet functions, filters, and dictionaries (see section C.4).

An overview of objects and class structure of S+WAVELETS is displayed in figure C.1.

C.1   1-D Transform and Analysis Objects

The classes of 1-D transform, decompose, and other analysis objects are listed in table C.1.

C.1.1   1-D Transforms

Most of the 1-D transforms inherit from the class `crystal.vector`. The only transform objects which inherit from `crystal.list` are

Class	Chap.	Created By	Shares Data Structure With
1-D Transform Objects			
dwt	2	dwt	crystal.vector[1]
wpt	7	wpt	crystal.vector[1]
wpt	7	best.basis	crystal.vector[1]
wp	7	.ptable	crystal.vector[1]
ptable	7	wp.table	crystal.vector[1]
ptable	7	cp.table	crystal.vector[1]
block.dct	8	block.dct	crystal.vector
block.cpt	8	block.cpt	crystal.vector
cpt	8	cpt	crystal.vector
cpt	8	best.basis	crystal.vector
cp	8	.ptable	crystal.vector
nd.dwt	11	nd.dwt	crystal.vector
atrous	11	atrous	crystal.list
rob.dwt	11	rob.dwt	crystal.list
crystal.vector			
crystal.list			
Decompose Objects			
mrd	2	mrd	decompose
mra	2	mra	decompose
decompose	7	decompose	decompose
Other 1-D Analysis Objects			
waveshrink	6	waveshrink	waveshrink
pcosts	7	pcosts	pcosts

[1] If boundary=infinite, then this class will inherit from crystal.list instead of crystal.vector.

TABLE C.1. Class structure for 1-D transform objects, decompose objects, and other 1-D analysis objects.

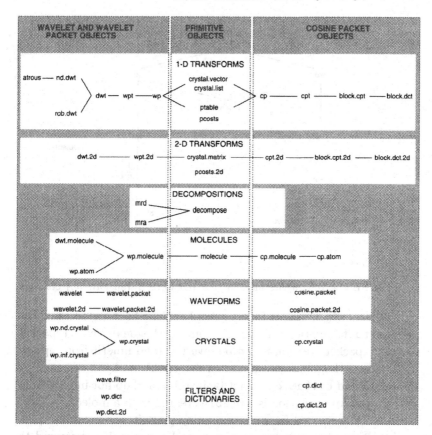

FIGURE C.1. An overview of the object-oriented design of S+WAVELETS. The horizontal boxes group the objects by type: 1-D transforms, 2-D transforms, etc. The objects are also categorized as primitive objects (middle vertical panel), objects specialized for wavelet and wavelet packet analysis (left vertical panel), and objects specialized for cosine packet analysis (right vertical panel). The lines indicate the inheritance hierarchy.

`atrous` and `rob.dwt`. Transforms which inherit from `wp` and have `boundary="infinite"` also inherit from `crystal.list`. The classes `crystal.vector` and `crystal.list` represent collections of crystals; objects of class `crystal.vector` are stored as vectors while objects of class `crystal.list` are stored as lists.

The inheritance hierarchy for 1-D transform objects is given in figure C.2. The root class is either `crystal.vector` or `crystal.list`. Objects which inherit from `wp` are wavelet packet transforms and objects which inherit from `cp` are cosine packet transforms. However,

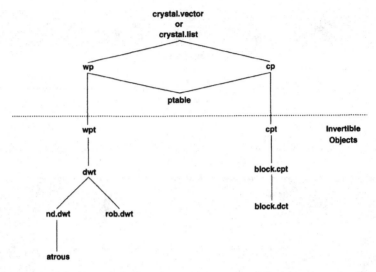

FIGURE C.2. Inheritance hierarchy for 1-D wavelet transform objects.

only objects which inherit from **wpt** and **cpt** can be inverted using the **reconstruct** function. In general, only orthogonal wavelet packet or cosine packet transform can be inverted (and inherit from **wpt** or **cpt**). Non-orthogonal transforms, such as **ptable** objects created by **wp.table** and **cp.table**, cannot be inverted and do not inherit from **wpt** or **cpt**. An exception is a **ptable** created from a molecule object (such as that created from the matching pursuit decomposition).

The classes **dwt**, **nd.dwt**, **rob.dwt**, and **atrous** all correspond to special types of wavelet packet transforms. Similarly, the classes **block.cpt** and **block.dct** correspond to special types of cosine packet transforms.

C.1.2 Decompose Objects

Decompose objects, which inherit from the **decompose** class, represent (additive) decompositions of data into separate components. They are created by applying the **decompose** function to a transform object. There are two special types of decomposition objects: multiresolution decomposition and multiresolution approximation created by the functions **mrd** and **mra**. See figure C.3 for an inheritance hierarchy for decompose objects. There is only one data structure for decompose objects.

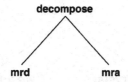

FIGURE C.3. Inheritance hierarchy for decompose objects.

C.1.3   Other 1-D Analysis Objects

Other 1-D wavelet analysis objects include packet cost tables and WaveShrink objects. Packet cost tables are generated automatically when a packet table is created and are computed using the **pcosts** generic function. WaveShrink objects are created by the **waveshrink** function.

C.2   Crystals, Molecules, and Atoms

Table C.2 lists the class structure for crystals, molecules, and atoms.

Class	Chap.	Created By	Shares Data Structure With
Crystals			
wp.crystal	2	wp.crystal	wp.crystal
wp.nd.crystal	2	wp.nd.crystal	wp.crystal
wp.inf.crystal	2	wp.inf.crystal	wp.inf.crystal
cp.crystal	8	cp.crystal	cp.crystal
Molecules and Atoms			
wp.molecule	10	matching.pursuit, top.atoms	molecule
cp.molecule	10	matching.pursuit, top.atoms	molecule
wp.atom	10	wp.atom	molecule
cp.atom	10	cp.atom	molecule
molecule			

TABLE C.2. Class structure for crystals, molecules, and atoms.

C.2.1   Crystals

Wavelet packet crystals (**wp.crystal**) and cosine packet crystals (**cp.crystal**) are the two basic types of crystals. A wavelet packet

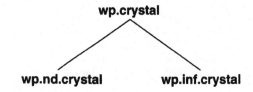

FIGURE C.4. Inheritance hierarchy for crystal objects.

crystal is a vector of wavelet packet coefficients for a given level j and oscillation b. A cosine packet crystal is a vector of cosine packet coefficients for a given level j and block b. They are called "crystals" because the coefficients are arranged on a lattice (in contrast to molecules).

The class hierarchy for crystals is shown in figure C.4. There are two special types of wavelet packet crystals: wp.inf.crystal and wp.nd.crystal. Crystals with wp.inf.crystal are created from wavelet packet transforms based on the "infinite" boundary conditions. These crystals have extra coefficients stored at the left and right boundaries, distinguishing them from the class wp.crystal.

C.2.2 Molecules and Atoms

Molecules form the basis for matching pursuit decompositions and decompositions created by the top.atoms function. Molecules are vectors of unordered wavelet packet or cosine packet coefficients. The coefficients can range over multiple levels j and oscillations or blocks b. Molecules are typically created using the matching.pursuit and top.atoms functions or through subscripting a crystal vector, crystal list, or crystal object.   Atoms are molecules of length 1, and

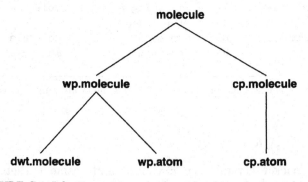

FIGURE C.5. Inheritance hierarchy for molecules and atoms objects.

are created using the [[subscript operator applied to a crystal or molecule.

The class hierarchy for molecules is shown in figure C.5. Analogously to crystals, there are wavelet packet molecules (wp.molecule) and cosine packet molecules (cp.molecule). Wavelet packet atoms (wp.atom) and cosine packet atoms (cp.atom) are special cases of molecules. Wavelet packet molecules further specialize into DWT molecules (dwt.molecule).

C.3   2-D Transform and Analysis Objects

The class structure for 2-D transform objects and other 2-D objects is given in table C.3.

Class	Chap.	Created By	Shares Data Structure With
2-D Transform Objects			
dwt.2d	3	dwt.2d	crystal.matrix
wpt.2d	9	wpt.2d, best.basis	crystal.matrix
block.dct.2d	9	block.dct.2d	crystal.matrix
block.cpt.2d	9	block.cpt.2d	crystal.matrix
cpt.2d	9	cpt.2d, best.basis	crystal.matrix
crystal.matrix			
Other 2-D Objects			
pcosts.2d	9	wp.costs.2d, cp.costs.2d	pcosts.2d
wp.crystal.2d	9	wp.crystal.2d	wp.crystal.2d
cp.crystal.2d	9	cp.crystal.2d	cp.crystal.2d

TABLE C.3. Class structure for 2-D transforms and analysis objects.

C.3.1   2-D Transform Objects

Figure C.6 displays the inheritance hierarchy for 2-D transform objects. The root class is crystal.matrix, which is a matrix which takes the same dimensions as the original image or matrix. There are two basic types of 2-D transforms: wavelet packet transforms (wpt.2d) and cosine packet transforms (cpt.2d). The wavelet packet

2-D transform specializes into a 2-D DWT and the cosine packet 2-D transform specializes into a block CPT and a block DCT transform.

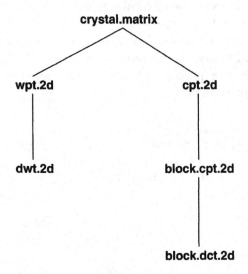

FIGURE C.6. Inheritance hierarchy for 2-D wavelet transform objects.

C.3.2   Other 2-D Objects

In addition to 2-D transforms, there are 2-D packet cost tables, created by the functions wp.costs.2d and cp.costs.2d. There are also 2-D wavelet packet and cosine packet crystals.

In contrast to the 1-D case, there is no analogue to the classes wp and cp: 2-D transforms are always orthogonal and invertible. Also, there is no analogue to the class ptable and there are no 2-D molecules or atoms.

C.4   Wavelet Functions, Filters, and Dictionaries

Table C.4 lists the class structure for class structure for wavelet functions, wavelet packet and cosine packet functions, wavelet filters, and dictionaries. Of these types of objects, only wavelet packet functions have a non-trivial inheritance hierarchy, which is shown in figure C.7.

Class	Chap.	Created By
Function Objects		
`wavelet`	2	`wavelet`
`wavelet.2d`	3	`wavelet.2d`
`wavelet.packet`	7	`wavelet.packet`
`cosine.packet`	8	`cosine.packet`
`wavelet.packet.2d`	9	`wavelet.packet.2d`
`cosine.packet.2d`	9	`cosine.packet.2d`
Filters and Dictionaries		
`wave.filter`	12	`wave.filter`
`wp.dict`	12	`wp.dict`
`cp.dict`	12	`cp.dict`
`wp.2d.dict`	12	`wp.2d.dict`
`cp.2d.dict`	12	`cp.2d.dict`

TABLE C.4. Class structure for wavelet functions, wavelet packet and cosine packet functions, wavelet filters, and dictionaries.

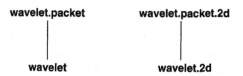

FIGURE C.7. Inheritance hierarchy for wavelet packet function objects. For these classes, there is no inheritance hierarchy and each object inherits from just one class.

Appendix D
S+WAVELETS Function List

This appendix contains lists of S+WAVELETS functions, grouped by their uses and tabulated with descriptions. Methods of generic functions are indicated by an asterisk (*).

Wavelet Convolutions and Filters

Topic	S+WAVELETS Function
Create Wavelet Filter	`wave.filter`
Wavelet Filter Methods	`plot*`, `print*`, `summary*`
Wavelet Filter Attributes	`is.dual`
	`is.high.pass`
	`is.low.pass`
	`is.orthogonal`
	`is.symmetric`
	`is.antisymmetric`
Plot Transfer Function	`transfer.plot*`
DWT Matrix Operators	`dwt.matrix`
	`wave.filter.matrix`
	`filter.matrix`
Convolution/Dilation Operator	`convdil`
Convolution and Down-Sampling Operator	`convdown`
Convolution and Up-Sampling Operator	`convup*`
Down/Up-Sampling Operators	`down.sample`
	`up.sample`
2-D Convolution and Down-Sampling Operators	`hor.convdown`
	`ver.convdown`
2-D Convolution and Up-Sampling Operators	`hor.convup`
	`ver.convup`
2-D Down/Up-Sampling Operators	`quad.down`
	`quad.up`

Cosine Packet (Local Cosine) Analysis

Topic	S+WAVELETS Function
Select Best Basis	best.basis*
Select Best Level	best.level*
Blocked Discrete Cosine Transform	block.dct print*
Blocked Cosine Packet Transform	block.cpt as.block.cpt print*
2-D Blocked DCT	block.dct.2d
2-D Blocked CPT	block.cpt.2d
Compute 2-D Cosine Packet Costs	cp.costs.2d
Cosine Packet Transform	cpt icpt print*
2-D Cosine Packet Transform	cpt.2d icpt.2d print*
Discrete Cosine Transform	dct dct.2d
Discrete Hartley Transform	dht
Tree Plot for Basis Selection	tree.plot
Signal Decomposition	decompose*
Inverse Transform Operator	reconstruct*
Create Cosine Packet Function	cosine.packet plot* print* summary*
Create 2-D Cosine Packet Function	cosine.packet.2d plot* print*
Support of Cosine Packet Function	support*
Create a Cosine Packet Atom	cp.atom plot* print* summary*
Create Cosine Packet Crystal	cp.crystal plot* print* summary*
Create 2-D Cosine Packet Crystal	cp.crystal.2d plot* print* summary*

Crystals

Topic	S+WAVELETS Function
Create Wavelet Crystal	wp.crystal
	plot* print* summary*
Create Infinite Wavelet Crystal	wp.inf.crystal
	left.pad left.pad<-
	right.pad right.pad<-
Create Non-Decimated Wavelet Crystal	wp.nd.crystal
Create 2-D Wavelet Packet Crystal	wp.crystal.2d
	plot* print* summary*
Create Cosine Packet Crystal	cp.crystal
	plot* print* summary*
Create 2-D Cosine Packet Crystal	cp.crystal.2d
	plot* print* summary*

Discrete Wavelet Transform Analysis

Topic	S+WAVELETS Function
Discrete Wavelet Transform	dwt idwt
	as.dwt
	plot* print*
2-D Discrete Wavelet Transform	dwt.2d idwt.2d
DWT Matrix Operator	dwt.matrix
Multiresolution Approximation	mra
	plot* print*
Multiresolution Decomposition	mrd
Non-decimated DWT	nd.dwt
	plot* print*
Robust Discrete Wavelet Transform	rob.dwt
	plot.rob.dwt
	print.rob.dwt
Select Atoms from Basis	top.atoms
Signal Decomposition	decompose*
Inverse Transform Operator	reconstruct*
Create Wavelet Function	wavelet
	print* summary*
Create 2-D Wavelet Function	wavelet.2d

High-Level Plots

Topic	S+WAVELETS Function
2-D Autocovariance or Autocorrelation Function	acf.2d
Bar Graph	barplot
Box Plot	boxplot*
Draw a Dot Chart	dotchart*
Exploratory Data Analysis Plots	eda.plot*
Energy Plot	energy.plot*
Plot Fourier Transform	fourier.transform.plot*
Grid Plot for a Packet Transform	pgrid.plot*
Basic Plot Function	plot*
Stack Plot	stack.plot*
Rescale Gray Levels	stretch*
Time-Frequency Plot	time.freq.plot*
Time-Scale Plot	time.scale.plot
Plot Transfer Function	transfer.plot*
Tree Plot for Basis Selection	tree.plot

Data Manipulation

Topic	S+WAVELETS Function
Coerce to Matrix	as.matrix*
Coerce to Vector	as.vector*
	as.crystal.vector
Math Group Method	Math*
Ops Group Method	Ops.*
Subscript and Assignment Methods	Subscript*
	*
	<-*
	*
	<-*
	$*
	$<-*

Miscellaneous

Topic	S+WAVELETS Function
Time-Frequency Plane Position	`bandwidth*`
	`center*`
	`freq.center*`
	`freq.bandwidth*`
Dyadic Refinement	`dyadic.refine`
	`phi.at.integers`
Testing Orthogonal Basis	`is.on.basis`
	`is.on.basis.2d`
Compute Maximum Level	`max.level`
Rescale Gray Levels	`stretch*`
Index Functions	`crystal.index`
	`node.index`
	`subband.index`
	`crystal.name.index`
	`get.parents.quad`
	`qnode.index`
	`quad.bound`
	`wavesubscript.to.index`
Crystal Names	`crystal.names*`
	`is.2d.name`
	`is.pretty.names`
	`icrystal.names`
Wavelet Filter Names	`wavelet.name*`
Supporting Functions	`binary.rep`
	`fast.reconstruct.wpt`
	`filter.length`
	`filter.start`
	`is.odd`
	`left.bell.fun`
	`plot.packet`
	`split.name.2d`
	`wave.filter.type`
	`wp.lengths`

Wavelet Molecules and Atoms

Topic	S+WAVELETS Function
Matching Pursuit Decomposition	matching.pursuit
Select Atoms from Basis	top.atoms
Molecule Objects	molecule.object
	plot* print* summary*
Create Wavelet Packet Atom	wp.atom
	plot* print* summary*
Create Cosine Packet Atom	cp.atom
	plot* print* summary*

Printing

Topic	S+WAVELETS Function
Basic Print Function (all objects)	print*
Compute a Summary	summary*

Smoothing Operations

Topic	S+WAVELETS Function
Wavelet Coefficient Shrinkage	shrink*
Estimate Spectrum with Wavelet Shrinkage	spec.wave
Wavelet Shrinkage Smoothing	waveshrink
	print* summary*

1-D Cosine Packet and Wavelet Packet Transforms

Topic	S+WAVELETS Function
A Trous Wavelet Transform	atrous
	print.atrous
Discrete Wavelet Transform	dwt idwt
	as.dwt
	plot* print*
Non-decimated Discrete Wavelet Transform	nd.dwt
	plot* print*
Robust Discrete Wavelet Transform	rob.dwt
	plot* print*
Wavelet Packet Table	wp.table as.ptable
Wavelet Transform	wpt iwpt
	print* summary*
Block Discrete Cosine Transform	block.dct
	print*
Block Cosine Packet Transform	block.cpt
	as.block.cpt
	print*
Cosine Packet Table	cp.table as.ptable
Cosine Packet Transform	cpt icpt
	summary*
Discrete Cosine Transform	dct dct.2d
Discrete Hartley Transform	dht
Select Best Basis	best.basis*
Select Best Level	best.level*
Select Atoms from Basis	top.atoms
Matching Pursuit Decomposition	matching.pursuit
Inverse Transform Operator	reconstruct*
Create Wavelet Function	wavelet
	print* summary*
Create Wavelet Packet Function	wavelet.packet
	print* summary*

2-D Wavelet and Cosine Transforms

Topic	S+WAVELETS Function
Select Best Basis	best.basis*
2-D Discrete Wavelet Transform	dwt.2d idwt.2d
2-D Block Discrete Cosine Transform	block.dct.2d
2-D Block Cosine Packet Transform	block.cpt.2d
Compute 2-D Cosine Packet Costs	cp.costs.2d
2-D Cosine Packet Transform	cpt.2d icpt.2d
	print.cpt.2d
Discrete Cosine Transform	dct dct.2d
Inverse Transform Operator	reconstruct*
	wp.costs.2d
2-D Wavelet Packet Transform	wpt.2d iwpt.2d
	print*

S+WAVELETS Signals, Images, and Data Sets

Topic	S+WAVELETS Function
Make Signals and Images	make.signal
	make.image
EEG Signal for an Evoked Response	eeg
Underwater Acoustic Signal of Ice Bangs	ice
Speech Signal of the Word "Had"	speech.had
Radar Glint Noise Signal	glint
NMR Spectra Signal	nmr1
MRI Image of Eve Riskin's Brain	brain
Cropped Digital Fingerprint Image	fingerprint
Digital Photograph of Ingrid Daubechies	daubechies
Gray Scale Image of a Telephone	phone
Scene Image with a Tank	tank
Gray Scale Image of a Fruit Bowl	fruits
Gray Scale Image of a Camera	camera
Gray Scale Image of a Bridge	bridge
Gray Scale Image of Lena	lena

Creating Wavelets, Wavelet Packets, and Cosine Packets

Topic	S+WAVELETS Function
Create Cosine Packet Function	`cosine.packet`
	`plot* print* summary*`
Create 2-D Cosine Packet Function	`cosine.packet.2d`
	`plot* print*`
Plot Fourier Transformation	`fourier.transform.plot*`
Plot Wavelet Functions	`plot.wavelet.packet`
Support of Wavelet/Cosine Packet Functions	`support*`
Create Wavelet Function	`wavelet`
	`print* summary*`
Create 2-D Wavelet Function	`wavelet.2d`
	`plot* print*`
Create Wavelet Packet Function	`wavelet.packet`
	`print* summary*`
Create 2-D Wavelet Packet Function	`wavelet.packet.2d`
	`print*`

Wavelet Classes and Objects

Topic	S+WAVELETS Function
Cosine Packet Crystal Objects	cp.crystal.object
Crystal List Objects	crystal.list.object
	as.crystal.list
	as.matrix.crystal.list
	as.vector.crystal.list
	plot* print* summary*
Crystal Matrix Objects	crystal.matrix.object
	as.matrix.crystal.matrix
	print* summary*
Crystal Vector Objects	crystal.vector.object
	plot* print* summary*
Signal Decomposition Object	decompose.object
	plot* print* summary*
Molecule Objects	molecule.object
	plot* print* summary*
DWT Molecule Object	dwt.molecule.object
Cosine Packet Molecule Object	cp.molecule.object
Wavelet Packet Molecule Object	wp.molecule.object
Cosine Packet Atom Object	cp.atom.object
Wavelet Packet Atom Object	wp.atom.object
2-D Pcosts Objects	pcosts.2d.object
	plot* print* summary*
Pcosts Objects	pcosts.object
	plot* print* summary*
Packet Table Object	ptable.object as.ptable
	plot* print* summary*
2-D Wavelet Packet Function Object	wavelet.packet.2d.object
2-D Wavelet Function Object	wavelet.2d.object
Wavelet Packet Object	wavelet.packet.object
Wavelet Function Object	wavelet.object
WaveShrink Object	waveshrink.object
Wavelet Packet Crystal Object	wp.crystal.object
	plot* print* summary*
Non-Decimated Wavelet	
Packet Crystal Object	wp.nd.crystal.object
Infinite Wavelet Packet	
Crystal Object	wp.inf.crystal.object

Wavelet Packet Analysis

Topic	S+WAVELETS Function
Wavelet Packet Table	wp.table as.ptable
Wavelet Transform	wpt iwpt
	print* summary
2-D Wavelet Packet Transform	wpt.2d iwpt.2d
	print*
Select Best Basis	best.basis*
Select Best Level	best.level*
Tree Plot for Packet Basis Selection	tree.plot
Signal Decomposition	decompose*
Create Wavelet Function	wavelet
	print* summary*
Create 2-D Wavelet Function	wavelet.2d
	plot* print*
Create Wavelet Packet Function	wavelet.packet
	print* summary*
Create 2-D Wavelet Packet Function	wavelet.packet.2d
	print*
Support of Wavelet Packet Function	support*
Compute 2-D Wavelet Packet Costs	wp.costs.2d
Create Wavelet Crystal	wp.crystal
	wp.inf.crystal
	wp.nd.crystal
	plot* print* summary*
Create Wavelet Packet Atom	wp.atom
	print* plot* summary*

References

[AH92]   Ali Akansu and Richard Haddad. *Multiresolution Signal Decomposition.* Academic Press, 1992.

[AWW92]   P. Auscher, G. Weiss, and M. V. Wickerhauser. Local sine and cosine bases of Coifman and Meyer and the construction of smooth wavelets. In Charles K. Chui, editor, *Wavelets: a tutorial in theory and applications*, pages 237–256. Academic Press, Inc., San Diego, CA, 1992.

[BA83]   P. J. Burt and E. H. Adelson. The Laplacian pyramid transforms for image coding. *IEEE Transactions on Communications*, 31:532–540, 1983.

[BBH93]   Jonathan N. Bradley, Christopher M. Brislawn, and Tom Hopper. The FBI wavelet/scale quantization standard for gray-scale fingerprint image compression. In *SPIE Proceedings, Visual Info. Process. II*, volume 1961, Orlando, FL, April 1993.

[BD95]   Jonathan B. Buckheit and David L. Donoho. Improved linear discrimination using time-frequency dictionaries. Technical report, Stanford University, 1995.

[BDGM94] Andrew G. Bruce, David L. Donoho, Hong-Ye Gao, and R. Douglas Martin. Denoising and robust nonlinear wavelet analysis. In *SPIE Proceedings, Wavelet Applications*, volume 2242, Orlando, FL, April 1994.

[BFOS84] L. Breiman, J. H. Friedman, R. A. Olshen, and C. J. Stone. *CART: Classification and Regression Tree.* Wadsworth, Belmont, CA, 1984.

[BG94] Andrew G. Bruce and Hong-Ye Gao. *S+WAVELETS Users Manual.* StatSci Division of MathSoft, Inc., 1700 Westlake Ave. N, Seattle, WA 98109-9891, 1994.

[BG95a] Andrew G. Bruce and Hong-Ye Gao. Understanding waveshrink: Variance and bias estimation. *Biometrika*, 1995. Accepted pending minor revisions.

[BG95b] Andrew G. Bruce and Hong-Ye Gao. Waveshrink: Shrinkage functions and thresholds. In Andrew F. Laine and Michael A. Unser, editors, *Wavelet Applications in Signal and Image Processing III*, volume 2569, pages 270–283, San Diego, CA, July 1995. SPIE.

[Bri92] Christopher M. Brislawn. Classification of symmetric wavelet transforms. Technical report, Los Alamos National Laboratory, Los Alamos, New Mexico, 87545, 1992.

[Bur95] Barbara Burke Hubbard. *The World According to Wavelets.* AK Peters, 1995.

[CD95] R. R. Coifman and David L. Donoho. Translation-invariant de-noising. In Anestis Antoniadis, editor, *Wavelets and Statistics*, pages 125–150. Springer-Verlag Lecture Notes, 1995.

[CDF92] A. Cohen, I. Daubechies, and J. C. Feauveau. Biorthogonal bases of compactly supported wavelets. *Comm. Pure Appl. Math.*, 45:485–560, 1992.

[CDV93] A. Cohen, I. Daubechies, and P. Vial. Wavelets on the interval and fast wavelet transforms. *Applied and Computational Harmonic Analysis*, 1:54–81, 1993.

[Chu92a]   C. K. Chui. *An introduction to wavelets*. Academic Press, Inc., San Diego, CA, 1992.

[Chu92b]   C. K. Chui, editor. *Wavelets: a tutorial in theory and applications*. Academic Press, Inc., San Diego, CA, 1992.

[CM91]   R. Coifman and Y. Meyer. Remarques sur l'analyse de Fourier à fenêtre. *C. R. Acad. Sci. Paris*, 312:259–261, 1991.

[CMQW90]   R. Coifman, Y. Meyer, S. Quake, and V. Wickerhauser. Signal processing and compression with wavelet packets. Technical report, Yale University, 1990.

[CMW92]   R. Coifman, Y. Meyer, and V. Wickerhauser. Wavelet analysis and signal processing. In *Wavelets and Their Applications*, pages 153–178. Jones and Bartlett Publishers, Boston, 1992.

[Cri93]   Criminal Justice Information Services. WSQ gray-scale fingerprint image compression specification. Technical Report IAFIS-IC-0110 (V2), Federal Bureau of Investigation, February 1993.

[CW92]   R. Coifman and V. Wickerhauser. Entropy-based algorithms for best basis selection. *IEEE Transactions on Information Theory*, 38(2):713–718, 1992.

[Dau92]   I. Daubechies. *Ten lectures on wavelets*. Society for industrial and applied mathematics, Philadelphia, PA, 1992.

[DJ93]   B. Delyon and A. Juditsky. Wavelet estimators, global error measures: Revisited. Technical Report Publication interne no. 782, IRISA-INRIA, 1993.

[DJ94]   David L. Donoho and Iain M. Johnstone. Ideal spatial adaptation via wavelet shrinkage. *Biometrika*, 81:425–455, 1994.

[DJ95]   David L. Donoho and Iain M. Johnstone. Adapting to unknown smoothness via wavelet shrinkage. *Journal of the American Statistical Association*, 90(432):1200–1224, December 1995.

[DJKP95]   D. Donoho, I. Johnstone, G. Kerkyacharian, and D. Picard. Wavelet shrinkage: Asymptopia? *Journal of the Royal Statistical Society, Series B*, 57:301–369, 1995. (with discussion).

[Don91]   David L. Donoho. Nonlinear solution of linear inverse problems by wavelet-vaguelette decomposition. Technical report, Department of Statistics, Stanford University, 1991.

[Don92]   David L. Donoho. Interpolating wavelet transforms. Technical report, Department of Statistics, Stanford University, October 1992.

[Don93a]   David L. Donoho. Nonlinear wavelet methods for recovery of signals, densities, and spectra from indirect and noisy data. In Ingrid Daubechies, editor, *Proceedings of the Symposia in Applied Mathematics*. American Mathematical Society, 1993.

[Don93b]   David L. Donoho. Smooth wavelet decomposition with blocky coefficient kernels. In Glen Webb and Larry Schumaker, editors, *Recent Advances in Wavelet Analysis*. Academic Press, 1993.

[Don95]   David L. Donoho. De-noising by soft thresholding. *IEEE Transactions on Information Theory*, 41(3):613–627, May 1995.

[Dut87]   P. Dutilleux. An implementation of the "algorithme à trous" to compute the wavelet transform. In J. M. Combes, A. Grossman, and Ph. Tchamitchian, editors, *Wavelets: Time-Frequency Methods and Phase Space*, pages 298–304. Springer-Verlag, 1987.

[Efr79]   B. Efron. Bootstrap methods: Another look at the jackknife. *The Annals of Statistics*, pages 1–26, 1979.

[EGK94]   Efi Eoufoula-Georgiou and Praveen Kumar, editors. *Wavelets in Geophysics*. Academic Press, Inc, 525 B Street, Suite 1900, San Diego, CA 92101-4495, 1994.

[Fla92]   Patrick Flandrin. Wavelet analysis and synthesis of fractional Brownian motion. *IEEE Transactions on Information Theory*, 38(2):910–917, 1992.

[FS81]   J. H. Friedman and W. Stuetzle. Projection pursuit regression. *Journal of the American Statistical Association*, 76:817–823, 1981.

[Gao93a]   Hong-Ye Gao. Choice of threshold for wavelet estimation of the log spectrum. Technical report, Stanford University, 1993. Under review at the Journal of Time Series Analysis.

[Gao93b]   Hong-Ye Gao. *Wavelet Estimation of Spectral Densities in Time Series Analysis*. PhD thesis, UC - Berkeley, 1993.

[Haa10]   A. Haar. Zur theorie der orthogonalen Funktionen-Systeme. *Math. Ann.*, 69:331–371, 1910.

[HBB92]   F. Hlawatsch and G. F. Bourdreaux-Bartels. Linear and quadratic time-frequency signal representations. *IEEE Signal Processing Magazine*, 9(2):21–67, April 1992.

[Hop94]   Tom Hopper. Compression of gray-scale fingerprint images. In *SPIE Proceedings, Wavelet Applications*, volume 2242, pages 180–187, Orlando, FL, April 1994.

[ISO91]   ISO/MPEG. Coding of moving pictures and associated audio for digital storage media at up to about 1.5 Mb/s. Draft International Standard DIS 11172, International Organization for Standardization, Dec 1991.

[JKP92]   I. Johnstone, G. Kerkyacharian, and D. Picard. Estimation d'une densité de probabilité par méthode d'ondelettes. *Comptes Rendus Acad. Sciences Paris (A)*, 315:211–216, 1992.

[JS94a]   Bjorn Jawerth and Wim Sweldens. An overview of wavelet based multiresolution analysis. *SIAM Review*, 36(3):377–412, September 1994.

[JS94b]   I. Johnstone and B. Silverman. Wavelet threshold estimators for data with correlated noise. Technical report,
Stanford University, 1994.

[Kol94]   Eric Kolaczyk. *WVD Solution of Inverse Problems.*
PhD thesis, Stanford University, 1994.

[LB78]   G. M. Ljung and G. E. P. Box. On a measure of lack of
fit in time series models. *Biometrika*, 66:66–72, 1978.

[LGO+95]   M. Lang, H. Guo, J. E. Odegard, C. S. Burrus, and
R. O. Wells. Nonlinear processing of a shift invariant
dwt for noise reduction. In *Wavelet Applications II*,
volume 2491, pages 640–651. SPIE, April 1995.

[LM95]   Jean-Marc Lina and Michel Mayrand. Image enhance-
ment with symmetric Daubechies wavelets. In *Proceedings of the SPIE 1995 Symposium, Wavelet Applications
in Signal and Image Processing*, San Diego, CA, July
1995.

[Mal86]   H. S. Malvar. Fast computation of the discrete cosine
transform through the fast Hartley transform. *Electronic Letters*, 22(7):352–353, March 1986.

[Mal89a]   Stéphane Mallat. Multifrequency channel decompositions of images and wavelet models. *IEEE Transactions
on Acoustics Speech and Signal Processing*, 37(12):2091–
2110, 1989.

[Mal89b]   Stéphane Mallat. A theory for multiresolution signal decomposition: the wavelet representation. *IEEE Transactions on Pattern Analysis and Machine Intelligence*,
11(7):674–693, 1989.

[Mey86]   Y. Meyer. Ondelettes, fonctions splines et analyses
graduées. Lectures, Univesity of Torino, Italy, 1986.

[Mey90]   Yves Meyer. *Ondelettes et opérateurs.* Hermann,
Éditeurs des sciences et des arts, 293 rue Lecourbe,
75015 Paris, 1990.

[Mey93]   Yves Meyer. *Wavelets: Algorithms and Applications.* SIAM, 3600 University City Science Center, Philadelphia, PA 19104-2688, 1993.

[MH92]   Stéphane Mallat and Wen Liang Hwang. Singularity detection and processing with wavelets. *IEEE Transactions on Information Theory,* 38(2):617–643, 1992.

[Mou92]   Pierre Moulin. Wavelets as a regularization technique for spectral density estimation. In *Proc. IEEE-Signal ProcessingSymposium on Time-Frequency and Time-Scale Analysis,* pages 73–76, October 1992.

[Mou93]   Pierre Moulin. Wavelet thresholding techniques for power spectrum estimation. Technical report, Bellcore, 1993. submitted to IEEE Trans. on Signal Processing.

[MW94]   Eric L. Miller and Alan S. Willsky. Wavelet transforms and multiscale estimation techniques for the solution of multisensor inverse problems. In *SPIE Proceedings, Wavelet Applications,* volume 2242, Orlando, FL, April 1994.

[MZ92]   Stéphane Mallat and Sifen Zhong. Characterization of signals from multiscale edges. *IEEE Transactions on Pattern Analysis and Machine Intelligence,* 14(7):710–732, 1992.

[MZ93]   Stéphane Mallat and Zhifeng Zhang. Matching pursuits with time frequency dictionaries. *IEEE Transactions on Signal Processing,* 41(12), December 1993.

[Nas95]   G. P. Nason. Wavelet function estimation using cross-validation. In Anestis Antoniadis and Georges Oppenheim, editors, *Wavelets and Statistics,* pages 261–280. Springer-Verlag, New York, 1995. Lecture Notes in Statistics.

[NS95]   G. P. Nason and B. W. Silverman. The stationary wavelet transform and some statistical applications. In Anestis Antoniadis and Georges Oppenheim, editors, *Wavelets and Statistics,* pages 281–300. Springer-Verlag, New York, 1995. Lecture Notes in Statistics.

[NvS95]   Michael H. Neumann and Rainer von Sachs. Wavelet thresholding: Beyond the gaussian I.I.D. situation. In Anestis Antoniadis and Georges Oppenheim, editors, *Wavelets and Statistics*, pages 301–329. Springer-Verlag, New York, 1995. Lecture Notes in Statistics.

[OP94]   Todd Ogden and Emanuel Parzen. Data dependent wavelet thresholding in nonparametric regression with change-point applications. Technical report, University of South Carolina, Columbia, SC, 29208, 1994.

[Per93]   D. B. Percival. An introduction to spectral analysis and wavelets. In *Proceedings of the Workshop, Applied Mathematical Tools in Metrology*, Torino, Italy, October 1993.

[PM95]   D. B. Percival and H. O. Mofjeld. Multiresolution and variance analysis using wavelets with application to sub-tidal coastal sea levels. *Journal of the American Statistical Association*, 1995. Under review.

[RBG94]   David L. Ragozin, Andrew G. Bruce, and Hong-Ye Gao. Non-smooth wavelets: Graphing functions unbounded on every interval. In *Recent Advances in Approximation Theory, Wavelets and Applications*. Kluwer, 1994.

[RV91]   Olivier Rioul and Martin Vetterli. Wavelets and signal processing. *IEEE Signal Processing Magazine*, pages 14–38, October 1991.

[RY90]   K. R. Rao and P. Yip. *Discrete Cosine Transform*. Academic Press, Inc., San Diego, CA, 1990.

[SC94]   Naoki Saito and Ronald Coifman. Local discriminant bases. In *SPIE Proceedings, Wavelet Applications*, volume 2303, San Diego, CA, July 1994.

[She92]   Mark J. Shensa. The discrete wavelet transform: Wedding the A Trous and Mallat algorithms. *IEEE Transactions on Signal Processing*, 40(10):2464–2482, 1992.

[SJBH85]   H. V. Sorenson, D. L. Jones, C. S. Burrus, and M. T. Heideman. On computing the discrete Hartley transfrom. *IEEE Trans. ASSP*, 33:1231–1245, 1985.

[SM65]   S. S. Shapiro and Wilk M. An analysis of variance test for normality (complete samples). *Biometrika*, 52:591–611, 1965.

[SMB94]   Jean-Luc Starck, Fionn Murtagh, and Albert Bijanoui. Multiresolution support applied to image filtering and restoration. Technical report, Observatoire de la Côte d'Azur, B. P. 229, F-06304 Nice Cedex 4, France, 1994.

[SN96]   Gilbert Strang and Truong Nguyen. *Wavelets and Filter Banks*. Wellesley-Cambridge Press, 1996.

[SP92]   W. Sweldens and R. Peissens. Quadrature formulae for the calculation of the wavelet decomposition. Technical report, K. U. Leuven, Belgium, 1992.

[Str89]   Gilbert Strang. Wavelets and dilation equations: A brief introduction. *SIAM Review*, 31(4):614–627, 1989.

[TK92]   A. H. Tewfik and M. Kim. Correlation structure of the discrete wavelet coefficients of fractal Brownian motion. *IEEE Transactions on Information Theory*, 38(2):904–909, 1992.

[Tri95]   K. Tribouley. Adaptive density estimation. In Anestis Antoniadis and Georges Oppenheim, editors, *Wavelets and Statistics*, pages 385–395. Springer-Verlag, New York, 1995. Lecture Notes in Statistics.

[US 95]   US Bureau of the Census. X-12-arima reference manual: Pre-release version 0.2. Technical report, Statistical Research Division, Room 3000-4, Washington, DC 20233-9100, April 1995.

[vSS94]   Rainer v. Sachs and Kai Schneider. Wavelet smoothing of evolutionary spectra by non-linear thresholding. Technical report, Universitat Kaiserslautern, Postfach 3049, D-67653 Kaiserslautern, Germany, Augest 1994.

[VV93]   P. F. Velleman and A. Y. Velleman. *Multirate Systems and Filter Banks*. Prentice Hall, 1993.

[VV95]   M. Vannucci and B. Vidakovic. Preventing the dirac disaster: Wavelet based density estimation. Discussion Paper 27, Duke University, 1995.

[Wan95]   Yazhen Wang. Jump and sharp cusp detection by wavelets. *Biometrika*, 82(2):385–97, 1995.

[Wic94a]   Mladen V. Wickerhauser. *Adapted Wavelet Analysis— from theory to software*. A. K. Peters, Ltd, 1994.

[Wic94b]   Mladen V. Wickerhauser. Wavelet approximations to Jacobians and the inversion of complicated maps. In *Proceedings of the SPIE: Wavelet Applications*, volume 2242, Orlando, FL, 1994. SPIE.

[WMP95]   A. T. Walden, E. J. McCoy, and D. B. Percival. Spectrum estimation by wavelet thresholding of multitaper estimators. Technical report, Imperial College of Science, Technology and Medicine, August 1995.

[XKS94]   Xiang-Gen Xia, C.-C Jay Kuo, and Bruce W. Suter. Improved Backus-Gilbert method for signal reconstruction with a wavelet model. In *SPIE Proceedings, Wavelet Applications*, volume 2242, Orlando, FL, April 1994.

Index